Beekeeping ᵀᴹ

Bringing Bee Keeping into the 21st Century to make Healthy Happy Bees

By Bill Summers

Northern Bee Books

Published in the United Kingdom by
Northern Bee Books,
Scout Bottom Farm,
Mytholmroyd,
West Yorkshire HX7 5JS
Tel: 01422 882751
Fax: 01422 886157
www.northernbeebooks.co.uk

ISBN 978-1-908904-69-0

Design and artwork, D&P Design and Print
Printed by Lightning Source (UK)

Beekeeping with ZEST™

Bringing Bee Keeping into the 21st Century to make Healthy Happy Bees

By Bill Summers

Northern Bee Books

Dedicated to William B. Summers.
Born 29/02/1920
Died 21/10/2014

CONTENTS

FOREWORD

Forward by Roy Pink who was responsible for the VIBEZ (Ventilated Intermediate Bee Entry Zone) concept for wood hives.

At first sight the Zest appears cumbersome and, contrary to the designer's background, an architectural eye sore. However, once passed all the coffin jokes, we found the Zest hive surprisingly easy to work with. Initial assumptions about using heavy frames were allayed when it was discovered the frames (being double brood depth) could be pivoted on one corner for inspection, the working height is perfect for manipulations, especially if you suffer with a bad back or are just a habitual leaner! Bill's determination and drive can be very addictive when working through a problem. His ability to 'stick with it' enabled him to finally arrive at the very impressive triple cross bar frame, even the bees toe the line and use the cells in each section appropriately! With an eye on economics, I believe a plastic frame is now being developed for cheaper production/recycling. But the main attraction for me has always been the top entrance and ventilation system. Preventing the flue effect and so reducing the need for extra heat via honey consumption in one fell swoop, is a master stroke that beekeepers can implement extremely easily within their own hives. Once beekeepers accept and become familiar with top ventilation (perhaps gained using a B.S. National VIBEZ conversion) the question of hive material arises. Bill makes a convincing argument against wood preferring a porous cement block that retains daytime heat, carrying this through the night within the hive. I think the question of what material (damp wood vs. insulating cement) has been answered. The question now is how it is best implemented. Bill has taken huge strides forward in developing the use of new materials within beekeeping. The improvement in the bee's confidence within their environment is displayed through faster expansion, no nosema , less varroa, good temperament and general good health throughout the colony. Bill has pushed the ZEST project along at such a pace. I'm definitely one beekeeper keen to see what this next edition will bring to the table for discussion.

Roy Pink
fish.r.pink@btinternet.com

PREFACE

The ZEST venture started with a stumbling attempt to make a bamboo bee hive frame for use in those countries that had plenty of it, mostly poor ones.

It widened into considering everything about beekeeping from the internal frames to the external envelope and even into beekeeper protection suits. A cursory inspection revealed that the planet wide bee keeping system and its artefacts were multiply flawed, but readily fixed by design.

Such a realisation brought with it the burden to speak, not with words written on the wind, but committed to paper for clarity and reference. It is a manifesto for bee health. It is most readily accepted by the beginner beekeeper who carries no historical baggage nor financial commitment.

There has been a recent doubling of beekeepers in the country caused by the ecologically minded seeking to assist honey bees in a better environment. This can be applauded. This book is to assist them and their bees.

The ZEST hive is democratic. Anyone can have one. It is cheap, appropriate and amenable to a more self-sufficient way of life. It is a living sustainable system, not a product. It can be entirely D.I.Y.

No one owns it. It can be free. Take it. Use it. Have fun.

Bill Summers

CHAPTER 1

INTRODUCTION

History shows that it is difficult to persuade the established order to accept revolutionary design change. It is greeted first with humour, followed by anger until finally it is seen to be self-evidently correct.

We accept that there is a crisis in honey bee health in which considerable numbers 3of bees die. Few have dared to suggest that wood hives, used for 160 years unchanged, are the cause. This book does so, explaining that traditional hives are physiologically unfit for honey bees, being cold and damp. We look to science to solve disease problems, but with a lack of success. An industry in bee palliatives has needlessly been born.

We look for strength in unity to our National Association. Its management structure is labyrinthine and undemocratic, yet its executive (against its own constitution clauses 3 and 5) sponsored (oxy-moronically) the policy of "bee friendly" insecticides,....for money. The membership finally managed to overcome the pressures placed upon it and revolted against the policy, yet without the benefit of seeing any resignations.

The National Association does not even now condemn the use of systemic neonicotinoids on the seeds of flowering crops that bees attend. It fence sits in the face of the dubious attractions and pressures of the crop protection industry. Neonicotinoids have now been banned in Europe for 2 years. The U.K. government chose not to support the vote to ban, but will reluctantly implement it anyway.

"Doing More with Less" is the ZEST ambition for technical design improvements that will prevail and sustain. Design/Science is the method to achieve the ambition. The ZEST hive and the associated range of ZEST equipment ruthlessly adopt these principles to enable bees to be healthy and productive with less energy.

This is not a book for the timid and faint hearted. If you believe in tradition and have high blood pressure it is better to look away now, because you are really not going to like this book.

It is a book for the experienced beekeeper grown weary of their bees dying of disease and of the cost of keeping them. It is amenable to "let alone" beekeeping.

It is a book for the beginner who is considering taking up beekeeping and does not wish to spend money buying last century's hives with their 28 design flaws. These are listed showing the ZEST response to those failures.

There are many excellent books about bees, beekeeping and how to do it. The number seems to grow weekly. They deal with the three aspects of keeping any creature and beekeeping is not an exception. They are Environment, Management and Breeding. There is, however, a void the size of the Grand Canyon in the available literature on the bee hive environment. It is the elephant in the apiary. It is for a technical design led approach to bee hives and their management that is less costly and more sustainable, healthier for bees and which can be built by anyone without obtaining a mortgage. What is needed is a simple, appropriate hive design that is based on a sound understanding of the physiology of bees and of modern materials. A ZEST hive is a cheaper and healthier hive environment for honeybees, being warm and dry rather than cold and damp. This is now proven with no winter deaths and less disease generally.

Also required is backup equipment of a nucleus box and honey warmer that can be made with a few D.I.Y. skills and which can reasonably be described as sustainable.

This treatise is intended to fill the void with the ZEST Hive and its range of supplementary equipment. They have, in use, improved the life and productivity of bees and beekeepers, raising the fertility of the eco system in which they are based and encouraged more people to keep bees. A virtuous circle can be established where let alone beekeeping is again possible. 21 ZEST hive colonies were taken into the winter of 2012/13 and 21 came out, all in good health and with surplus stores. 3 of those were queen less, but after the mating conditions of the 2012 summer this was not surprising. Similar results occurred during the winter of 2013/2014. 3 queens failed out of 34 colonies before Christmas, but there were no further losses due to disease in the remaining 31. They are also proving resilient to spite and swarming, with superseding being preferred. These are the intended consequences of habitation redesign for the honeybee. In retrospect honeybees appear to have been let down by humanity for a century and a half.

The ZEST hive design has also exhibited a profound, but unintended consequence. They appear to be functionally varroa free. This is a bold statement, but is based upon there being no dead varroa in the floor debris of a number of

ZEST hives, whose owners are listed here below. They have agreed to be called by anyone wishing to confirm the zero varroa count.

Julie Challoner 01963 363810

Sue Ferguson 01747 850420

Pete Metcalf 07990 721524

Dave Durrant (ZEST brother in arms) first noticed, researched and raised the matter, confirming it as likely to be so by reference to the following beekeepers:-

Erik Österlund http://www.elgon.se/index-eng.htm

Michael Bush http://www.bushfarms.com/

Dee Lusby http://www.beverlybees.com/dee-lusby-organic-beekeeper/

Hive debris from a ZEST hive shows no fallen varroa.

The varroa mite's exponential growth in a honeybee colony is determined by the time it has in the bee's pupating cells to itself mature. A sufficient reduction in pupation time reduces its prospects to one of exponential decline. This seems to have happened. We are unsure of the mechanism, but our best guess so far is as follows:-

1. The honeycomb in the ZEST hive is naturally drawn out as in the wild, but on a controlling plastic lattice moveable frame. This natural cell size averages about 4.9mm whereas the wax foundation sheets generally used in traditional hives is 5.4mm. This naturally smaller cell is likely to reduce the bee's pupation period.

2. The ZEST hive is warm and dry. Due to its insulation, thermal mass and top bee entry/ventilation it is more easily thermo-regulated by the bees to 35 degrees. This is likely to speed the biological process, again reducing the pupation period of the bees.

The Varroa mite originated in the Asian honeybee Apis Cerana. It is located on the eastern side of the Himalayan Mountains, which had formed a barrier to its expansion. Man's activities moved it into the western honeybee (Apis Mellifera) where there was no natural resistance to Varroa's exponential expansion in numbers. This was alleged to be caused by the longer pupation period of the western honeybee, which allowed the Varroa mite an extended maturation period in the pupating cell. It was assumed that the Asian honey bee had evolved a 2 day shorter pupation period to achieve stasis in varroa numbers. Perhaps this is not the case though.

Drones have a longer pupation period than workers or queens in both species so the varroa mite naturally prefers (by 10 or 12 times) to lay eggs in the drone cells of both species. Despite this preference it was not just the drones in Apis Cerana that had apparently evolved a shorter pupation period, but workers and queens as well, which points away from the evolution of a shorter pupation period. The honeybee grub/pupa gestation period appears to vary upon ambient and brood nest temperatures rather than to have evolved to be shorter.

If hotter conditions and smaller cells do encourage shorter pupation periods in honeybees then it is logical that colder conditions such as winter and larger cells will do the opposite. This would give the Varroa an advantage that it was formally unaccustomed to in Asia and may explain why the spring is particularly prone to seeing Varroa in large numbers after a winter of perhaps longer pupation periods in worker cells.

The development of the ZEST Hive began in 2008 and has proven to be a better environment for housing the biological system of honeybees. This is not to forget the other 2 aspects of management and breeding of honeybees that needs addressing. These are also dealt with here, but with less vigour than environment which seems to key to improving the lot of honeybees and subsequently of their management and breeding.

The original ambition was to design a hive suitable for beekeeping in developing countries where resources are few, but which did include concrete blocks, discarded metal roof sheets, bamboo for making frames, and fishing line to hold them together. Examples were built and tested as proof of concept. Modifications were made with reservations about the fishing line, which stretched. This was discarded and replaced with brass split pin paper fasteners normally used for holding bank statement together.

The ZEST hive design then focussed on developed (cold) country use incorporating loose laid ultra-lightweight insulated blocks for walls, floors and roofs. They accommodated double depth British Standard National width brood frames fixed one above the other with ply strips joining them at the sides. Conversion for beekeepers from wood hives to insulated block ones could then be a natural progression using their existing B.S. frames and bees. Larger frames such as Langstroths can also be used by simply increasing the internal width of the ZEST by 90mm. to receive them. A greater width of frame would reduce the ratio of external wall to volume.

The natural habitat of choice for honeybees is a cave or a large hollow tree that has high thermal capacity. This moderates the temperature as with a storage heater (or cooler). We do not seem to have improved on this habitat of choice for the bees, but have made matters worse for them with thin walled wood hives. They are not appropriate to their designated task of enabling bees their health and of beekeepers their profit.

The single and double walled traditional hives are the housing equivalent of a cardboard box under Waterloo Bridge. It is furthermore made from shockingly expensive cardboard with high maintenance needs and with the functional over-design of Heath Robinson on steroids. The Victorians bequeathed them to us, but beekeepers have clung tenaciously to them despite their obvious failings.

The ambient temperature in which traditional wood hives exist is constantly changing over the seasons, from day to night and over the day caused by weather. The bees need to constantly thermo-regulate the brood temperature at 35°C if the brood is to survive and thrive. This not only stresses the bees, but is costly in honey, both to heat the hive and to collect water to cool it.

Bees can survive if they are wet or cold, but not both. Thermo-regulation results in the constant variation in the relative humidity in the hive causing condensation on cold surfaces. This condensation is the same as that which we would experience from a hot shower in a cold bathroom. Condensation is particularly noticeable in winter when the temperature difference between the brood nest and the outside can be 40°C. It is a measure of bee resilience and a true wonder that any bees at all survive the winter in the wood boxes that we give them to live in.

In order to "design out" these fundamental flaws in traditional wood hives the ZEST hive concept for hive design and management was initiated. One was for use in ambient weather conditions. The other was heated for comparison. The heated hive did shockingly well when compared to the cold ZEST hive. There may even have been some "drift" to the warm hive. Roy Pink remained committed to the experiment throughout the 2011/12 season and who also accepted responsibility for testing the JUMBO ZEST hive which contained 4 colonies. He became particularly interested in the effect on Varroa infestation in ZEST hives heated during the winter.

Hunter gathering of honey was replaced by the keeping of honeybees in straw skeps hundreds of years ago. In mid Victorian times the Rev. Langstroth, while not in church, invented the movable frame hive and patented it. He placed them in a wood brood box based on a champagne crate. The first was a sensible concept, because we could then manage the bees without killing them. The champagne crate was not so smart, since it bore none of the natural thermo-regulation capabilities of a cave or large hollow tree. We are no longer restrained by the need to drink champagne in order to have somewhere to keep bees. With the ZEST we have designed something better. It is time that we did, but first some broad philosophy about design against which any design can be measured.

It can be too easy to make heroes, but one that can be readily accepted is Buckminster Fuller whose thoughts follow here.

The word "design" is generally understood as being what it looks like and/or how it works. Buckminster Fuller had insights into the design process which are worth reporting before moving on. As an architect and a general system theorist he was ahead of his time on the matter of design, who considered it to be the driving force for the advance of civilisation, technology and the winning of wars at sea by "The Great Pirates". Those "Pirates" who could do "MORE WITH LESS" won the battles. The battle of Trafalgar was won by the British fleet being able to fire their guns every 4 minutes against 12 minutes for the French and Spanish, whose ships were quickly turned from gun platforms into burning wrecks. The British Empire was built on the subsequent mastery of the oceans. On such apparently small matters are the fortunes of nations and empires determined.

Buckminster Fuller spoke of Ecology half a century ago and was aware of the unity and utility of everything which he called simply "Universe".

To do MORE WITH LESS in energy terms is the fundamental drive of life where the efficient use of energy is paramount to survival, reproduction and evolution. Evolution is driven by consciousness's pursuit of pleasure and avoidance of pain.

Buckminster Fuller considered that any invention or design that would survive and thrive would have to do MORE WITH LESS. An example includes the radio valve, which was superseded by the transistor, which in turn was superseded by the microchip. All are gates through which an electrical signal passes in one direction, but each did **more with less** material and energy. Another example is of a stacking, polypropylene chair which superseded hand or machined wood ones. Further darker examples include the AK47, atomic weapons and the stealth bomber.

The concept of "Design/Science" was also initiated by Buckminster Fuller. Design and Science are subjects that are usually seen at different ends of a school curriculum with Humanities in the middle and Sport somewhere outside in the rain. The school approach tends to give us design "as it looks". It could be reasonably accused of trivialising design in which design is no longer inventing, designing, constructing and using made objects, but exhibiting good taste, found in the department stores across the planet, not on its battlefields...... nor in its apiaries.

Buckminster Fuller said that we have to make up our minds to make sense or make money, if we want to be designers. To make sense, designs need to be made from the things we see around us that are perhaps already being used for something else, are deemed waste and have a low embodied energy.

Buckminster Fuller's understanding of Science is "taking something apart to understand it" and of Design as "putting something together" to use it. Design and Science were therefore as inseparable as two sides of the same coin. Without Science there can be no Design and without Design there is no reason for Science. This leads seamlessly on to "Design Method". What is the process of Design if not a simple "how it looks" selection of colour, proportion and texture?

There are five strands to the process that occurs when designing something that is beyond a mere visual experience. These strands are **Objective/ Research/ Design/ Construction/ Feedback**. This is not a linear process, but a circular one over time. Research, when carried out will perhaps lead to a restating of the Objective. The Design process may indicate a fresh Research strand to be pursued. Feedback is not something that happens at the end of the design or construction process, but permeates the whole.

Design may result in an object, but may also be just a system change.

CHAPTER 2

OBJECTIVE (THE BRIEF)

This book is not an attempt to teach beekeeping, but to say why the ZEST hive is a good choice to make for keeping bees. It remains a work in progress. When you are down the pub or at the bee association you can amaze your friends, comrades and the barmaid with your results. She may even want to take you home.

They may all be less than totally believing about the honey you have collected, how disease free and easy your bees are to manage with their good temper, and at such little cost. They may begin to see you as a Walter Mitty fantasist. Tell them to get their own ZEST and prove it for themselves.

As a beginner you will have been taught on the usual wood hives hearing the lessons that your teachers heard. As a thoughtful beekeeper you will realize the limitations. This will have been good experience as bad experience is the best you will get. It teaches you more. Better however to use someone else's bad experience like that of the author. Beginners usually lose swarms from bad tempered colonies in their first year (I did), the remainder of which get disease in the winter (mine did). Many give up. Do not allow this to happen. The world needs ZEST beekeepers to save the planet, its ecosystem, put honey on toast and fruit in the bowl for the kids.

Considerable sums of tax-payer's money are spent on matters such as sequencing the genome of the bacteria that causes EFB. The Waggle Tail Dance and the hunt for a "hygienic" bee take up more. European money is tipped into a project that will monitor bees via a satellite link to the beekeepers living room. It is a puzzle to know just what the benefits of all this expenditure is except to keep scientists employed doing their Ph.d's. and paying off their mortgages. While all this has been going on, my bees had been dying quietly and in quantity from Nosema during the winters. This is the principle cause of winter losses in traditional hives, because Nosema flourishes in a cold, damp environment. We all know this. Money could be spent more effectively on the down to earth question of better hive design.

Diseases

Thought to be caused by a virus, but latterly thought to be a fungus! to be exacerbated by cold and damp hive conditions in winter. The removal of Fumidil B from the market which acted as a palliative for the disease may lead to more Nosema in our bee stocks. Prevention rather than cure has now been made compulsory. The solution is to make the bee environment warm and dry rather than cold and damp.

2. CCD (Colony Collapse Disorder). Some of the latest thinking on the cause is expressed in the BFA Bulletin of June 2010.

It suggests that Nosema combines with a fungus to collapse the colony.

SAN DIEGO, CA – May 25, 2010 -- New research from the United States Department of Agriculture (USDA) identifies a new potential cause for Colony Collapse Disorder in honeybees. A group of pathogens including a fungus and family of viruses may be working together to cause the decline. Scientists report their results today at the 110th General Meeting of the American Society for Microbiology in San Diego.

There might be a synergism between two very different pathogens, says Jay Evans of the USDA Agricultural Research Service, a researcher on the study. When they show up together there is a significant correlation with colony collapse disorder

To better understand the cause of these collapses, in early 2007 Evans and his colleagues collected bees from both healthy and declining colonies across the country but primarily from California and Florida where most of the commercial pollination activity takes place. They have screened these samples and similar samples from each year since then for both known and novel pathogens.

They found a slightly higher incidence of a fungal pathogen known as Nosema ceranae in sick colonies, but it was not statistically significant until they began pairing it with other pathogens.

Levels of the fungus were slightly higher in sick colonies, but the presence of that fungus and 2 or 3 RNA viruses from the family Dicistroviridae is a pretty strong predictor of collapse, says Evans.

Nosema are transferred between bees via the fecal-oral route. When a bee initially ingests the microbes and they get to the mid-gut, they harpoon themselves into the gut wall and live inside the epithelial cells there. Evans believes that the slightly higher numbers of the fungus somehow compromise the gut wall and allow the viruses to overwhelm the bees. In colonies with higher Nosema numbers they found virus levels to be 2-3 times greater than healthy colonies.

While this is a working theory and they are still in the discovery phase looking for new pathogens, Evans and his colleagues are also actively looking for a way to boost bee defences against Nosema.

A way to protect against Nosema might be the key for now, says Evans.

3. Acarine – Seen in the spring and also described as K-wing. It is caused by a tracheal mite.

   See Page 7 of "The Beekeepers Quarterly" No.95 March 2009 where John McMullan Ph.D of Trinity College Dublin writes on Page 7.

   "In sub-tropical regions of the world tracheal-mite infestation levels can be very high, but the colonies will not normally die. Only in regions with cool winters do deaths normally occur. It has only recently been identified (McMullan and Brown, Exp. Appl. Acoral., in press) that the mechanism causing death in tracheal-mite-infested colonies is the inability of the colonies to thermo-regulate".

   Reducing the bees difficulty in thermo-regulation in a cold climate may be the way forward, which may enable bees to live with the endemic tracheal mite and not be affected.

4. Varroa – While this is not a disease, but a mite, it is the vector for enough diseases to collapse the hive. In a climate that is cold in winter, into which varroa has moved as an alien species, its numbers appear to have increased from mathematical stability to exponential expansion. Its numbers are determined by it having enough time in the pupating bee cells to itself mature.

   Reducing this pupation period may be the way forward

5. European Foul Brood – This is caused by a bacteria living in the larva gut and which eats the larval food. The larva starves and dies. It decays, giving the characteristic foul smell of putrefaction. It seemed to be a disease found more in heather areas which are on acid soil, but the Random Apiary Surveys (RAS) have disproved this. It is unknown for certain whether cold and damp exacerbates the disease, but its incidence increases in summers that are cold and damp as in the summer of 2012.

   The cure when found is:-

   a) Destruction of bees with petrol, burning of frames and scorching of boxes.

This is reserved for serious cases.

b) Shook swarm in which only the adult bees are kept and put onto new frames. The old frames are burnt. Success rate is about 80%.

6. American Foul Brood. Rare, but deadly. This has only one possible cure. Unlike EFB it is one of destruction and burning and has, perhaps as a consequence, been virtually eliminated in the U.K.

This treatise is primarily about bee health and how to improve it with a hive design that is less cold and damp. The disease problems caused by cold and damp is perceived as being caused by the hive architecture. While dealing with that on bee health grounds it became apparent that traditional hives are a costly Victorian anachronism best replaced by the ZEST principles of habitation design.

Cross top ventilation and bee entry in the ZEST directly removes damp, carbon dioxide laden air while maintaining the brood cluster temperature below. There is no flue/stack effect cooling the colony, because there is no air inlet at low level. The ZEST insulated external hive envelope also reduces the temperature difference between the inside and outside of the hive allowing the bees to thermo-regulate the brood more easily........their prime ambition for colony survival.

Readers can draw their own conclusions on this thesis. They may choose to switch to ZEST Hives, whether as a beginner or with experience, or to remain wedded to the existing technology and wood designs. While there continues to be no laws against cruelty to bees the choice remains yours...... No pressure.

This is the statement, not about designing a better mousetrap, but a better beehive. This is the brief, but what constitutes better? It is doing **"More with Less"** using **Design/Science**. How can we get more honey.....and by implication...... more pollination with less energy and materials, than we do at present? Once this broad objective is stated we cannot simply do a lot of pointing and instruct the bees to comply and provide. We need to gain a deep empathy with the bees to understand what makes them want to naturally co-operate in our ambition. The relationship is one of mutual co-operation. Our task is to provide good quality housing, a health service and food when there is famine. There is a school of thought in "natural beekeeping" that this includes letting bees succumb to "natural diseases" when we "unnatural humans" could prevent them doing so. We seem to have trouble accepting that we humans are also natural and deserving of space in the Universe together with our ambitions for bees, which are entirely supportive.

What honey bees want from the beekeeper, as far as possible is to be **Warm and Dry** and:
a) Be disease, pest and parasite free.
b) Behave naturally.
c) Fill the Universe with their kind.
d) Nurture their young at 35°C.
e) Have adequate accessible food and water.
f)   Have a hive secure against weather and predators.
g) Be kept busy doing productive things such as drawing wax.
h) Not be overcrowded, but to have a compact brood nest.
i)   Be free of insecticides
j)   Have a prolific Queen, mainly found in her first year.
k) Not to have pollen or honey blocks preventing the Queen laying eggs.

If these wants are fulfilled the bees are unlikely to swarm, but will supersede. I invite you to accept that they think like us. Why would they want to vacate a fine dwelling that is easy to run and healthy.

Most of these "bee wants" are not readily met with thin walled wood hives managed in the way that we do. If these wants are met the bees are quieter, harder working and better tempered as indeed we are when our needs are easily met.

What beekeepers want from the bees is to be:
a) Productive, making a surplus of honey for harvesting at minimum cost.
b) Good tempered.
c) Disease free.
d) Pollinate crops.
e) Able to brag about the honey they harvested and how many colonies survived the winter.
f) Not be stung, injured, tired or made uncomfortable.

CHAPTER 3

3.0 RESEARCH

Changes in scientific knowledge and the useful design technology that research brings can have profound effects, leading to re-consideration of existing designs and systems. This happens very quickly in times of war, but slowly or not at all in times of peace.

The introduction of modern building technology, when used for building pastiche, has not resulted in a total reconsideration of building design in the light of the demise of fossil fuel. Tinkering at the edges takes place rather than starting again as we should. Old, inappropriate, architectural forms are retained despite their cost. Introducing essential, small scale renewable energy devices onto a new Georgian pastiche dwelling is an impossible expectation without a radical reassessment of the importance of "good taste". We remain conservative, tired and lazy as the natural default position.

And so it is with hive design. The reusable, moveable frame hive was invented and used without the benefit of electricity. Before electricity, extraction was by crushing and draining then with centrifugal extractors. Motors were added to save the manpower to use the extractor.

The original method of crushing, warming and straining that were employed during the time of keeping bees in skeps was not reverted to, because of the perceived value of drawn comb being reusable.

The technology of thermostatically controlled and insulated honey warmers came much later and remains an add-on rather than a replacement to centrifugal extraction and frame reuse.

We are now advised that comb re-use is not wise, since it carries disease. Such a perception alone calls for a fundamental reconsideration of honey gathering, beehive design and management. The beekeeping methods that we currently use are reaching the point of systemic collapse due to the collected overburden of inefficiencies. The system is over designed, because of a series of incremental changes. A radical re-design is overdue that will "Do More with Less".

RESEARCH 3.1.
EXISTING HIVES

A. WOOD HIVE DEFICIENCIES AND THE ZEST RESPONSE

Do we use wood hives out of habit, despite their failures, when we would prefer to give them up in favour of a better design, if only we could just get out of bed to do it?

Long beekeeping experience identifies the following significant number of inadequacies in the traditional wood hive, which the new ZEST Hive has been designed to overcome:-

1. The thermo-regulation of wood hives is a constant and heavy duty for the bees. It needs to be carried out by the bees on an hourly, daily and monthly basis as the ambient temperature varies. A colony of bees can collect an average of 600kgs.of nectar in a year which is inverted into 300kgs.of honey. Honey and nectar is used by the bees to warm the colony directly, but also indirectly by using it to fly and collect water to cool the hive. Honey used in thermo-regulation of the colony is entirely wasted.

The ZEST is insulated all round, with insulated building blocks that also act to give a thermal "capacity" to moderate the speed of temperature fluctuation. This assists the bees in maintaining an even colony temperature without using much honey or nectar.

2. Traditional wood hives, with an entrance at the bottom and ventilation at the top, develop a "stack effect" in which an updraft is caused, as in a chimney flue. The temperature difference between the brood nest and outside in the winter can be 40C. The stack effect is severe with such a temperature difference when it would be best not present at all. In the summer the stack effect is much less with its higher outside, ambient temperature in which the temperature difference between inside and out is much smaller. This is when the stack effect is needed to cool an overheated hive, but does not do so due to the smaller temperature difference.

The ZEST has top, cross, trickle ventilation under the roof, which maintains the temperature of the colony, allowing the gentle exhausting of moist, carbon dioxide laden air at high level. There is no stack effect, because there is no air inlet at low level.

3. The stack effect found in traditional wood hives prevents the bees from warming up and expanding the colony in the early spring. They tend to build up on the nectar flow rather than collect it for the rent. Stimulation feeding is needed every early spring.

The ZEST colony does not dwindle so quickly in a severe winter, because it is warm and dry. It is all ready to go with the flow as the days lengthen.

4. The bees in a traditional hive try to reduce the cooling stack effect at high level by filling gaps up with propolis. This causes condensation to form on the underside of the cover board and drips on the bees.

The ZEST method of top, cross, trickle ventilation is enough to remove any high humidity air and prevent condensation forming.

5. The cost of traditional wood hives is prohibitive.

The cost of ZEST hives are about a third of traditional wood ones when the frame area for what each provides is compared.

6. Traditional wood hives are bespoke like pieces of furniture, which are then left out in the rain.

The ZEST hives are made from readily available materials, often being recycled such as sheet metal for the roofs and blocks for the foundation.

7. Traditional wood hives need regular maintenance and repair.

The ZEST hives need none, being robust.

8. Traditional wood hives are prone to be stolen, woodpecker damage, overturning by a hungry badger or by a fierce wind.

The ZEST is prone to none of these, being robust. It may not even need a stock proof fence, but this is unproven.

9. Traditional wood hives vary in size over the year and require storage for items of equipment during winter.

The ZEST hive equipment is stored in the hive all year. Only the volume for the bees varies.

10. The floor board entrance of traditional wood hives means that the bees tend to dump nectar in the brood chamber as they pass up through it. They are in a hurry to offload it and get back out. They fill cells that the queen should be laying in. She is prevented from doing so. This is the worst management failure. The bees will move some of the nectar up in the night, but not always and why make them double handle it anyway?

The ZEST arrangement of top entrances and no queen excluders allows the bee's easier access to where the nectar and pollen is stored.

11. The low entrance of traditional wood hives means that the weeds have to be kept down in front of the hive.

The ZEST entrances at high level do not need clearing of weeds.

12. The traditional (vertical) wood hive can only contain 1 colony and only be increased in size in box sized chunks.

The ZEST management plan shown in Drawing 8A and 8B can contain 2 colonies over winter and be restricted or expanded on a frame by frame basis.

13. The traditional wood hives have different sized frames in the same hive. They are not interchangeable. Honey in the deep brood chamber cannot be moved above the queen excluder (into shallower supers) to free up space for the queen to lay eggs in. Brood congestion is the principle cause of swarming.

The ZEST has one frame size.

14. The use of queen excluders in traditional wood hives is a complication. Denmark does not use queen excluders in its traditional hives. They encourage swarming.

The use of vertical queen excluders was originally tested during 2010 in a ZEST which did not work well. Further design and testing may improve the use of excluders, but this does not appear likely. Perhaps it is better so.

15. The re-use of drawn out wax frames carries disease between colonies. When clean wax is provided for the queen to lay in she is usually very quick to do so, showing her preference.

The ZEST does not re-use comb. Wax is remade naturally by the bees and is always fresh and new for the queen.

16. Traditional wood hive boxes when full of honey can break strong men's backs and the Health and Safety Regulations for load lifting. Less robust persons are driven to give up or not take up beekeeping. If designed today they would be unlawful.

The maximum load that requires manipulation in a ZEST is a 4kgs.frame of honey. All are accessible and at the right height for standing upright.

17. The volume of a traditional wood hive cannot be easily adjusted to enable the colony to keep itself warm. The choices are a brood box, a brood and a half or two brood boxes. The volume needed for the brood chamber changes throughout the year, but the difficulty of changing it in small increments means that the volume cannot match the colonies needs close enough.

The ZEST can be adjusted in size on a frame by frame basis using insulated division boards at each end of the colony. The colony space can be made compact, but not overcrowded, by fine adjustment.

18. There is a tendency with the traditional wood hives for the bee keeper to want to only bring away a whole box of honey after it has been cleared of bees. Empty frames and both granulated and unripe honey can be present in the box being taken away.

The honey in a ZEST is taken away on a frame by frame assessed basis or even on a third of a frame basis. Each frame is taken only when the honey is ready.

19. Taking honey from a traditional wood hive requires 3 visits. To put on the Porter bee escapes, then to return and take the box of honey after 2 nights clearing. (It can be completely robbed by the bees if there is the smallest access for them to do so). The final visit is to return the wet frames for cleaning up by the bees.

For small amounts the ZEST needs only one visit to harvest the honey. It can be brushed off of bees, cut out into a nylon mesh bag in a bucket and removed. The frame is returned immediately to the hive.

Alternatively for larger amounts the bees can be brushed off the frames,

placed on a sheet and stood up leaning in the back of a car to be carried away for taking.

Another alternative is for the frames to be put into the ZEST nucleus box, with the bees and Canadian (traffic cone) escapes in the floor. The box is placed over the feed holes in the ZEST roof. The bees clear back down to the queen in 2 nights allowing the box of honey to be taken away free of bees. A few drones are usually left to brush off. Health and Safety load lift limits are likely to be exceeded.

20. When boxes of honey are taken away from the apiary at least a third of the weight is box and frames. Honey extractable from a super box by a centrifugal extractor varies between 25% and 90%. The remainder must be returned to the bees. A lot of superfluous carrying goes on.

The ZEST extraction method allows 100% extraction. The expensive centrifugal extractor is replaced by hanging nylon bags over 20 litre plastic tubs and a honey warmer for the last of the honey.......which is needed anyway to turn granulated honey in jars into runny honey in jars.

The ZEST honey is not "spun out", but "run out" which can be sold as such at a premium. Perhaps honey is not affected by a good battering on a stainless steel wall, but this is unlikely.

21. The bees in the traditional wood hive tend to stick all the parts together with propolis and there are many parts. Cracking the boxes and the frames within them, and then replacing them one above the other kills bees. The hive needs to be cleaned up by the bees. Every entry to a traditional hive is said to cost a jar of honey. This manipulation shortens the temper of the bees.

The ZEST has no boxes to crack and then replace. The frames do not sit one above the other. Only the roof comes off. No bees need to be killed. They are generally sublime.

22. We enter traditional wood hive in a manner that impersonates a bear. As a result the temper of bees in a traditional wood hive is poor. A smoker is usually needed.

The temper of bees in a ZEST is better. The reason is not obvious, but is likely to be the relatively stealthy opening up of the hive. The roof is lifted. No bees need to die. It may also be for the reason that the bee's thermo regulation duties are less. A smoker is not needed.

23. The 6 frame nucleus box of traditional wood hives is generally not big enough to overwinter successfully.

The 6 frame ZEST nucleus box being twice the frame depth is large enough to overwinter successfully. It is also designed not to have a stack effect with trickle top cross ventilation and bee access as in the ZEST hive.

24. The traditional nucleus box (with a traditional wood hive) cannot be used to clear bees from frames of honey.

The ZEST nucleus box can be used to clear bees from honey by deploying them above the ZEST hive and with Canadian escapes in the nucleus floor between them. (See 19 above)

25. The traditional wood hive needs spare frames for various eventualities and expansion.

The ZEST can be run with exactly the right number of frames, since they are constantly re-used.

26. The traditional wood hives have to be to British Standards.

The ZEST system as a DIY system allows a hive of any size and shape within the ZEST hive design principles.

27. The bees in a traditional wood hive are prone to attach (in a crowded hive) the side bars of the frames to the inside of the wood box.

The ZEST insulated block hive walls are powdery and the bees cannot attach comb to it successfully. If done at all it is slight and is easily detached.

28. Traditional hives are not naturally amenable to the shook swarm (from beginning of June until beginning of August) method of management, because of the cost of the replacement wax foundation every year and the culling of brood.

See Drawing 8A and 8B for ZEST management options of nucleus and artificial swarm making.

The ZEST hive has the ability to hold 1 large colony of the equivalent volume of 5 B.S. National Brood boxes or 2 colonies of half the size, which is the

intended overwintering position. This gives options in the following spring to which bee keepers are unaccustomed. It provides insurance against winter losses together with a surplus of colonies for sale in the spring.

B. THE BRITISH STANDARD / ZEST FRAME COMPARISONS

The British Standard frame disadvantages

They are:
1. Made with complicated joints which take time to assemble.
2. Not a DIY make item.
3. Not suitable for use in the third world where machine technology is not available
4. Often twisted in the face plane causing comb indiscipline.
5. Does not always sit vertical in the hive under gravity.
6. They are often stuck down hard with propolis.
7. Releasing them can result in broken end lugs.
8. Frame bottom bars can stick to the frame below and break when removed.
9. Uses unneeded foundation wax.
10. Expensive.
11. Designed for undesirable re-use of comb.
12. 12 Hoffman frames in width equal 11 Standard spacer frames!
13. Wax foundation faces distance apart can vary between 30mm and 45mm.
14. Surface, rather than point supported at each lug end causing them to hang unevenly.
15. Needs to be wire reinforced for extraction, which deters the bees.
16. Needs plastic spacer ends which must be removed for extraction of honey.

The ZEST frame advantages.

They are:
1. In the absence of wax foundation bees will build their own natural comb vertical under gravity. It may twist and turn on plan, but it will be vertical. The ZEST gravity frame establishes a build line on plan and uses this to ensure as accurate comb building as possible.

2. Point supported at each end allowing them to swing like pendulums until they settle under gravity. All exactly the same.
3. The bees are allowed to act naturally by working upside down drawing wax comb rather than off a face of wax foundation.
4. The comb is not reused in another hive which could pass on diseases.
5. The bamboo frame is appropriate for third world use and being collapsible can be made and packed up tight in a box for sale. It is just shaken out and used.
6. Having side bars they overcome the comb indiscipline and other problems of just top bar hives.
7. Bees like to be kept busy making wax.
8. Frame support runners not required.
9. Centrifugal honey extractor not required.

Three types of ZEST gravity frame are deployed, each suitable for use in its context.

a. The collapsible bamboo frames (See Drawing 9)

This is appropriate for use in the third world where the raw material often grows like a weed.

This is the original prototype, made and tested in a B.S. National hive at the end of 2008 and was held together with fishing line. It proved the concept of frames without foundation.

The final bamboo frame is now made out of an 8 foot long cane, cut into its lengths and holes drilled on a pillar drill with a jig. The junctions are held together with brass coated split paper fasteners and is (of course) collapsible. To make and assemble 200 of these took about 10 hours.

The bees draw the spine of the comb from the lowest part of the circular horizontal canes.

b. The ZEST stapled wood frames (See Drawing 10)

The second frame type is made from machined wood strips stapled together on a jig at the junctions and with wax starter strip grooves under the horizontal bars. The wood for 600 frames cost £300 and to machine it a further £300. They took about 40 hours to assemble and fit wax starter strips.

c. The ZEST plastic frame.

The third and final frame type is plastic as shown and is a bought item in boxes of 12 from www.thezesthive.com

The inventive part of this design, which is patented, is the deployment of

T-Bars, the tails of which are the starting point and guide for the spine of the natural honeycomb.

RESEARCH 3.2.
EXISTING WISDOM

A.MIGRATING BEEKEEPING

The only discernible advantage of wood hives over the ZEST is that they can be easily bundled up and carried off to another site where they can collect honey and provide pollination services. Is it wise though, to do so?

In the U.S.A. migration is carried out three times during a year in big trucks on long roads between Florida, Washington and California. Accidents happen and diseases are spread.

It is not finally proven that this trucking about is bad for bees, but there is strong circumstantial evidence to that effect. It is certainly true that diseases are spread rapidly by migrating bees and is likely to be exacerbated by the stress of doing so.

The primary reason for migrating beekeeping is the capital cost of the hives. An income from pollination can be obtained by moving them 3 times in a year so the capital cost of the hives can be recovered.

If an economical hive such as the ZEST hive was available, it would be more expensive to move bees than to provide permanent colonies at each of the migration sites. A ZEST is about a third of the cost for the same comb face area than for traditional hives.

Diversity under planting of bee forage for when the pollination crop was not in flower would resolve the negative effects on bee health of living in a short term monoculture crop. Almond and citrus groves could be under planted with a variety of bee forage plants for before and after the pollination crop flowers, but not during. It would be a simple matter to take the honey once a year. The ZEST hives would be non-migratory not only for the financial reason that they are too cheap to bother moving, but because it cannot be picked up and moved without falling apart. A big deterrent. If a cheap, immovable hive such as the ZEST hive was designated as a national standard this would raise beekeepers confidence in it.

This would bring positive results for bees, beekeepers and fruit farmers requiring pollination services. If the farmer provided the ZEST hives he might even be able to sell the honey taking rights rather than having to pay for pollination services.

B.SWARM CONTROL

Hive management includes the concept of "swarm control". Swarms are seen as a bad thing, because it reduces the honey take as bees are lost. Swarming is caused by c), g), h) and k) in the "bees want" list namely to:

c) Fill the Universe with their kind.

g) Be kept busy doing productive things such as drawing wax.

h) Not be overcrowded, but to have a compact brood nest.

k) Not to have pollen or honey blocks preventing the Queen laying eggs.

In a world where bee colony numbers are naturally declining, the concept of "swarm control" has become a misnomer. We now need an artificial "swarm inducing" system in place that does not reduce the honey take and becomes a seamless part of a positive management technique.

C.THE SEARCH FOR THE HOLY GRAIL "HYGIENIC" BEE.

Bees are already hygienic, but this ill-informed ambition gives the impression that they are not. As a consequence it seems that people with power, but with little knowledge of bees have sought to alleviate the negative consequences of these "unhygienic" bees by seeking a ban on the marketing of honey from cells that have contained bee larvae. It is the most cleansed environment in the hive.

Considerable time and effort has been, and continues to be, devoted to the task of breeding "hygienic" bees that can dispel varroa naturally, but there is nothing to show for it. It always seems to be just one more research grant away. Selective breeding of animals has value when contained by fences, but when they cannot be, it has none. Artificial selection based on the perceived "hygienic" behaviour of clearing out brood killed by freezing is not logical.

The criteria for selection by breeding is the speed at which the bees clear out killed brood, but the varroa mite does not wish to kill the brood (like any sensible parasite), but lives on it. The speed at which the bees clear out the dead brood is not a measure of "hygienic" behaviour. It is a measure of all the variable factors associated with one colony over another, such as strength, season, position of frame, need for frame space, age profile and other diseases present.

Let us assume for a moment that it was possible to breed a "hygienic" bee that cleared out pupa infected with varroa. It would need to be bred in an isolated apiary of which there are few......and mostly abroad. How could this "hygienic" bee then be introduced into the outside world of beekeeping?

If simply released, the first cross colonies will be infected with "unhygienic" qualities brought by "unhygienic" wild drones, which will die out without treatment. A colony cannot be half alive or dead. A colony must be treated or it must not. If it is treated natural selection cannot take over. If it is not treated it will die. Natural selection requires the survival of the fittest. There is no sign of that happening and every "cross" will reduce the "hygienic" quality, if indeed it ever truly existed.

Would other bee colonies therefore be culled before the "hygienic" bee was introduced into an area? To ensure that there were no other "unhygienic" feral colony drones to "cross out" any "hygienic" qualities. Would there be a two year moratorium on honeybees in that area? The logistics of such a "hygienic" bee introduction program would be immense and bound to fail. Recalcitrant beekeepers would need to be shot at dawn together with their bees.

If a "hygienic" honeybee was successfully introduced, genetic diversity would be reduced to that of the first "hygienic" bee which would then fill the world with its kind. Such loss of genetic diversity may be a serious matter that would leave the bee population vulnerable to passing pathogens. We may then look back with fondness on a lost world, albeit infected with varroa.

D.THE WAGGLE TAIL DANCE.

The generally accepted reason for the waggle tail dance may be wrong. It is obvious that bees dance, since we have all seen it. There must be a reason, but if not to give direction and distance to forage then what could it be for?

The dance may **not be the direct cause of the effect** of bees foraging where they would not otherwise do so. The dance may be to attract a crowd in a noisy and dark hive by vibrating comb rather like a town crier's bell. The dance may be the precursor to giving directions to a group of dance followers by antenna touching (or some other method).

If so, antenna touching must be described as a language in which abstract concepts required for giving directions is possible, as in humans.

Human language operates at two levels. "One to one" and "One to many". Bees may do the same. "One to one" may involve 2 bees touching antenna. This is seen often, but about what, we can only speculate if not associated with the waggle dance.

The dance, by attracting followers "one to many" would remove the element of "Chinese whispers" inherent in passing on information on a "one to one" hierarchical basis. The dancer tells all those followers who report for forage duty by antenna touching to give directions.

Human language uses the two way process of thought-speech-hearing-thought and requires a range of sounds between speech and hearing to convey meaning. The English language has 46 phonetic sounds and is sufficient for the language of Shakespeare. Bees would probably need less.

A bee language conveyed by antenna touching may operate at the ultrasound level, but there is little evidence that this exists in honey bees, although it can be concealed in a tube such as an antenna and be undetectable. Radio waves are another option, but would require a "crystal set" structure involving vibrating hairs as a receiver and a mineral as a transducer. (def: it converts one form of energy into another, eg mechanical to electrical and vice versa). A transmitter at a distance (perhaps the pore plates) would require an electrical discharge of considerable power which does not seem available to bees. Language may be sent in another way such as *in extremis* direct "download" of experience by antenna touching. It may even be a language based on smell receipt and detection. Bees can differentiate many more than 46 (phonetic) smells.

We know that bee antenna have hairs on their antenna, as humans have in their ears for hearing. In humans these act as transducers to receive impulses which are transmitted to the brain and interpreted as sound. The anatomy of bee antennae includes pore plates, hairs and a segmented exoskeleton. This structure is surely capable of conveying **abstract** meaning by an antenna touching sign language.

I am neither alone nor part of a majority among beekeepers in suggesting that the conventional reason for the waggle tail dance is wrong. An article by Caroline Williams (features editor) in New Scientist of the 19/09/2009 also raised doubts. I have a suspicion that Karl von Frisch's theory is a post decision rationalisation that has become accepted wisdom. Bee researchers certainly spend a lot of valuable time on it.

It is a conventional belief among beekeepers that the waggle tail dance is a "run", but in reality it is a "stand". See "The Buzz About Bees" by Jurgen Tautz. On page 95 Mr.Tautz states that dancers only move one or two steps to get a better grip. It is an optical illusion that it is a "run". A dancing bee has an audience of dance followers whose antenna are held rigidly at attention facing the dancer.

Antenna touching occurs on the return runs. On Page 108 and 109 Mr. Tautz raises his own "surprising anomalies" in the logic of the waggle tail dance. He shows courage and a startling disregard for conventional wisdom. Good for him. If you want to know what they are.....buy the book. On page 99 he makes the point that it is not necessary for a bee to know the distance to a forage source, only the direction. If known the bee can fly till she finds it by olfaction. He says that distance knowledge is a luxury. Darwin excludes luxury.

On Page 106 Mr.Tautz states:

"While the message about the location of the food source is most likely received over the antenna an important link in the procedure is missing. How do interested bees find dancers on the busy, crowded and dark dance floor?"

That link may commence by broadcasting like a town crier. It may be no more than a plea to listen by vibrating the comb. The more efficient the vibration is, the greater the audience is that seems to turn up. The most efficient dances are on open cells where a proper grip can be obtained.

On page 101 Mr Tautz refers to the experiment where dancer bees have previously flown down a deception tunnel which deceives the bee into thinking she has flown much farther than she has. He concludes that:

"It explained and settled the decade-long controversy about the waggle dance, in which it was disputed whether or not the recruited bees followed the information coded in the waggle tail dance. The tunnel enabled one to produce bees that made errors, visiting feeding sites 6m. from the hive, but in their dance signalling a distance 30 times longer. Searching recruits were not found flying around where the dancing bee really came from, but in an area much further away where there was nothing of interest. **Information from the dance is (therefore) used"**.

This may not be the correct conclusion. It may mean that the information has indeed been transferred, but not necessarily by the waggle tail dance.

Harvey J. Gold said on the subject of Mathematical Modelling of Biological Systems.

"The result of a mathematical development should be continuously checked against one's own intuition about what constitutes reasonable biological behaviour. When such a check reveals disagreement, then the following possibilities must be considered:

a)   A mistake has been made in the formal mathematical development.

b)   The starting assumptions are incorrect and/or constitute a too drastic oversimplification.

c) *One's own intuition about the biological field is inadequately developed.*

d) *A penetrating new principle has been discovered.*

Does the waggle tail dance constitute reasonable biological behaviour to direct dance followers to forage? It is inaccurate, sequentially variable, unseen in the dark and done on horizontal surfaces, negating its association with gravity. I am less than totally believing.

The organisation seen in a colony of bees cannot reasonably be expected to be carried out by the waggle tail dance, pheromones, olfaction and instinct alone. A bee language capable of conveying **abstract** thoughts would explain much about unexplained bee behaviour.

Honey bees have had 30 million years to develop language which would improve efficiency and therefore survivability. If the laws of the physical universe and Darwinian rules do not exclude the possibility it is highly likely, if not a certainty, that it has evolved.

In support of the theory of bee language there appears to be a collective memory (knowledge handed down) in a bee colony such as bees building queen cells. They have never seen or made one before. To say that this is just instinct is to use the word as the dumping ground for the "not understood". Bees have good memory and knowledge may be passed down. This again suggests language. Googling the key words olfaction, electrostatics, electret, chemosensillae to see that an electrostatic charge is present in the pore plates on the bees antenna. This is from a paper by Eric H. Erickson Jr. published in 1982. He speculates that it is present to attract molecules to improve olfaction. It may do this, but its presence may also bring other uses such as language transmission.

Googling "Dynamic range compression in the honeybee auditory system toward waggle dance sounds" also reveals a paper that infers the possibility of a language capable of conveying and receiving abstract concepts, yet then struggles to fit its findings to the waggle dance **as** *a language.*

It seems that further analysis of the tactile behaviours of bee antennae on returning from foraging and dancing needs to be video analysed frame by frame in order to identify any patterns that suggest a language. Is the antenna touching after the waggle tail dance consistent, yet different from the touching that occurs at other times and other circumstances may be the first question.

To determine whether a language based on antenna touching does indeed exist in honey bees, evidence may lay in the antenna touching ritual being the same between different bees when they are conveying the same information to each other. If it is seen that a dancing bee conducts the same antenna touching ritual with more than one dance follower it should be assumed that the same ritual carries the same information regarding direction and distance to forage.

A similar assumption can be made regarding the antenna touching that occurs between a guard bee and a bee seeking entry to the hive.

Jurgen Tautz in his first class book draws similarities with bees and mammal behaviour giving them nominal mammal status. If a language is part of the bee repertoire he could have been bolder.

Less mammal, more Homo Sapiens.

Written by Bill Summers

Edited by Mrs. Susan Ferguson (BSc. Hons. Zoology)

Mr. Stuart Ferguson (BSc. Chemistry)

Beekeeping with ZEST

RESEARCH 3.3.

A. VARROA

ZEST hives (August 2014) proved to be functionally free of varroa being not present in the hive debris. The following text was written prior to this being known, but is repeated here unchanged as a record of previous speculation.

Bees have a natural inclination to clean a hive of any extraneous matter, undesirable pests and parasites. Varroa mites are recovered from hive debris that appears to have suffered damage such as a dented carapace that could only have been caused by the bees.

If any animal's environment is difficult in some way (such as cold and/or damp habitat) it is unlikely that it attends to its cleaning duties as well as it might. Humans living in a house with walls running with water, draughts and is cold are less likely to be bothered than with a house with central heating that is in sound condition. Rats may even be accepted by humans as a comparatively minor problem as varroa may be, by bees, in their environmentally unfriendly hives.

It is a certainty that a healthy hive environment will allow the bees to spend more time and effort in destroying varroa. The only question is to what degree this takes place.

The Varroa mite originated in the Asian honeybee *Apis Cerana*. It was only located on the eastern side of the Himalayan Mountains, which had formed a barrier to its expansion. Man's activities moved it into the much more productive western honeybee (*Apis Mellifera*) where there was no natural resistance to Varroa's exponential expansion in numbers. This was alleged to be caused by the longer pupation period of the western honeybee, which allowed the Varroa mite an extended maturation period in the pupating cell. It was assumed that the Asian honey bee had evolved a 2 day shorter pupation period to achieve stasis in varroa numbers.

Drones have a longer pupation period than workers or queens in both species so the varroa mite naturally prefers (by 10 or 12 times) to lay eggs in the drone cells of both species. Despite this preference it was not just the drones in *Apis Cerana* that had apparently evolved a shorter pupation period, but the whole genus, which points away from the evolution of an evolved reduced pupation period. The honeybee grub/pupa gestation period appears to vary upon ambient and brood nest temperatures rather than to have evolved to be shorter.

We have also been forcing honey bees to draw out their honeycomb from wax foundation which has cells embossed on it at 5.4mm. This is an unnatural worker sized cell which would normally be an average of 4.9mm. When the bees are determined to make drone cells they override the 5.4mm, increasing it yet further to contain drones. It seems that the egg to hatch period is extended by both ambient temperature and cell size. The smaller the cell the quicker it is to hatch. In high temperature conditions, queen cells can hatch in 14 rather than 16 or 17 days and so it must be with workers and drones, reducing the Varroa's time to mature in the cells.

If hotter conditions and smaller cells encourage shorter egg to hatch periods in honeybees then it is logical that colder conditions and larger cells will do the opposite. This would give the Varroa an advantage that it was formally unaccustomed to and may explain why the spring is particularly prone to seeing Varroa in large numbers after a winter of perhaps longer pupation periods in worker cells.

The way forward in suppressing varroa would appear not to be breeding a hygienic bee, but warmer hives and natural (smaller cell size) comb.

In discussion with Tony Wright and Roy Pink working on the ZEST hive concept in North Devon it was agreed that they would independently test the concept of the ZEST insulated hive and compare it with a ZEST insulated and heated version. This took place during the 2010 season.

As a result of this experiment we began discussing some circumstantial evidence that points to a possible new understanding of the varroa/bee life cycle and which may lead to a technical method of both disrupting the varroa life cycle while giving the more obvious advantages to bees of a heated (and dry) environment.

If an alien species is moved into an environment it is unlikely to live in stasis with it. It will either expand or contract (and die out, but we do not hear of these). Examples of expanding species are the rabbit, cane toad and indeed British humanity in Australia. The absence of natural negative influences on those creatures in the new environment allowed them to survive, thrive and even to play better cricket.

And so it may be with the varroa mite.

The conventional wisdom regarding the varroa's success in colonising *Apis Mellifera* is that the egg/hatch period of the Mellifera drone was 2 days longer than in Asian bee, upon which the mite originated. Asian bee drones had apparently evolved to live in stasis with the varroa mite by reducing its egg/hatch period by 2 days. The theory went on to say that the 2 extra days of *Apis Mellifera* gave the mite the edge so that a rising population of mites became the norm on *Apis Mellifera*. So far so good, but now consider these facts as well:

1. The short period of very hot weather of 2010 disrupted 2 queen breeders in North Dorset where bred queen cells had all hatched at less than 15 days rather than 17. One of these breeders was the author.

2. African bees are able to live with varroa. (See "Just a Mite" on Page 86 of the Beekeepers Quarterly issue June 2010).

3. Bees in Asia live with varroa apparently due to the reduced pupation period of the drone in which its eggs are laid, but the reduced period occurs in the queen and workers as well. Why should this be?

4. There is a bee keeper who lives in Toulouse that claims to have honeybees immune to varroa and sells them out as such.

5. There remain feral colonies in (centrally heated) house roofs that show no signs of succumbing to varroa.

6. The three Mediterranean countries of Spain, Italy and Greece now have a rising honey bee population while the countries north of there have a declining one.

The common feature of these six locations is hot, hot, hot, hot, hot and hot.

Small overwintered nukes were found to be entirely free of the varroa marker of deformed wing virus, perhaps because they could not sustain any brood throughout the winter. The varroa may have died of old age before they could reproduce.

It may be that very cold and very hot are not good for varroa but for different reasons. The first may be, because there is no brood and the latter because the bee's pupation period is reduced in very hot conditions.

We may have a perfect climate for varroa in Northern Europe. Perhaps heating even in the summer is the way forward for varroa control in summer. Heating in summer would move the colony south into an African climate where varroa does not overwhelm the colonies.

Both the cold and heated ZEST hives have not, so far, exhibited any presence of nosema and after 4 years of testing remains so. The test has only been in small numbers so far. Time and numbers will prove it or not.

If the egg lay to hatch period is indeed reduced from the accepted norm in very hot ambient conditions in all bee species then it is likely that a cold winter ambient temperature will increase that period. An extended pupation period of say 2 days longer for pupation than the norm during cold ambient temperatures would give the varroa mite an opportunity for exponential growth in its numbers throughout the winter. They are certainly seen in larger numbers in the spring than in the autumn. This can be put down to being fewer bees to see the mites on in the spring, but this may not be the whole story.

Perhaps the Asian bee's 2 day shorter pupation period for the queen, drone and worker may be due not to evolution, but the higher temperature in Asian countries which speeds the bee breeding cycle, enabling it to live in stasis with the varroa mite.

It was discovered during 2011 that while the winter heated hive did better than the unheated ones the varroa mite count was also raised. This does not negate the theory that varroa are ambient temperature sensitive for the reasons stated. The raised temperature of the heated hive may have been not high enough or consistent enough.

B. THE SEARCH FOR A VARROA ATTRACTANT

This is notable in the U.K. for its absence. The American Agricultural Ministry has done some work which appears to support the possibility of a trap method for varroa. They claim to have found 2 chemicals that the mite is attracted to and have made a trap they claim that works. If true this is good news. They are not at present keen to tell anyone what these chemicals are until patents are in place, but if it does work as intended we will have a treatment that is entirely acceptable to all of us. A varroa attractor in a trap rather than a poison has a lot of style.

With money becoming available in the UK for bee disease research little thought appears to have been given to the prospect of making a varroa trap upon similar lines to a cockroach trap, which uses an attractant. The advantage of an attractant to trap the varroa mite (rather than a varroaicide) is that it is a harmless to bees and their products. The mite cannot develop immunity to an attractant without modifying its behaviour to become harmless.........i.e. by not being attracted.

Ross Conrad in his book "Natural Beekeeping" published in 2007 by the "Chelsea Green Publishing" in Vermont tells us that Methyl Palmitate is the Varroa pheromone and it Googles as such. He did not have a viable trap and gave up on the idea after he suggested in humour that he should patent it. Methyl Palmitate may yet have its day with a viable trap. Assuming that a trap can be found that works the task becomes finding and proving an attractant. There are substances other than Methyl Palmitate that have possibilities.

Peanut butter may work, but a successful attractant (if not varroa pheromone) is more likely to be based upon the chemical difference between a worker and drone grub immediately prior to sealing. It is known that the varroa mite has a 10-12 times greater propensity to lay its eggs in a drone cell than a worker. All creatures have a sense of taste/smell and the varroa mite is likely to use this to find the drone grubs to lay eggs in its cell. It has evolved to do so, because the drone is sealed for 2 days longer than the worker and this increased drone pupation period allows the varroa to gain a reproductive advantage, increasing its numbers more up to the point (and beyond) of collapsing the colony. The Asian bee, from where the varroa mite originated, has allegedly evolved to reduce its drone pupation period by two days and can therefore live in stasis with it.

So what is the chemical difference of a drone grub from a worker grub that the varroa mother finds so attractive and can the difference be determined with the use of Gas Chromatography chemical analysis with a view to making the chemical difference for use as an attractant. Has it already been done?

If you Google the following key words........ larval food / Varroa Destructor / cell invasion / semiochemical the second item on the Google page is worth a read. It does seem as though larval food is an attractant to varroa,

If varroa destructor is attracted to larval food and has a 10-12 times propensity to lay eggs in a drone cell it is reasonable to postulate that larval food (Royal Jelly) is greater in a drone grub than a worker.

If true this goes against the received wisdom that the drones are fed larval food for the same period as the workers before moving to a mixture of pollen and honey. There is a persuasive symmetry where both the queen and drone are fed only on royal jelly. It may be that the worker is the odd one out by being fed pollen and honey when still a larva.

If true, this would also explain why the varroa mite overwhelmingly prefers to travel on nurse bees which produce the royal jelly. They are meals on wheels.

The chemical difference of drone and worker larval food is likely to be seminal to making an effective chemical attractant for varroa. The best I could find on the matter was in a book by Mark L. Winston called The Biology of the Honey Bee on Pages 61 and 62 which gives a list of the chemicals in drone food, any of which may be the attractant to the varroa mite. Conversely any chemicals in the worker larva that are not in the drone may be a repellent to them. I quote from Mr. Winstons book (page 62) as follows:

The older drone larva receives more diverse proteins than the worker larvae. *Food of older drone larvae contains* **more** *carbohydrates, riboflavin and folic acid and* **less** *thianin, biotin, pantothenic acid, choline, pyridoxine, protein, fat, ash and niacin. These may form the basis for an essence of drone varroa attractant.*

C.THE VARROA LIFE CYCLE.

The received wisdom is that the varroa mite lives for about 27 days, but can live much longer (up to 5 months or even a year) in the winter for (presumably) the purpose of carrying it through the winter. This longer period is suspicious and is hard to prove. Were all varroa marked in some way and then found 5 months later? Furthermore the varroa mite evolved in a hotter Asian climate where brood is always present. How did the mite suddenly gain the ability to live for many months (5 times longer) when it came here where brood may not be present all year?

Roy Pink in smug mode with 4kgs.honey.

Roy found the following by Jeanne Pierre Chapleau dated March 2002 (revised March 2003) and which supports our own theories on the matter regarding ambient temperature and its effect on varroa breeding success.

The important difference in the global results obtained in 2000 (29.2% more varroa mites) and 2001 (37% less varroa mites) for sub group AV suggest a confirmation of the negative thermal influence assumed in the 2000 trials. In 2000, all of the anti-varroa bottom boards were operated with the bottom opened while in 2001, with the exception of the YBO group, the bottom boards were operated with the bottom closed. To our knowledge, this is the only operational factor that was systematically different between the 2000 and 2001 trials. The results strongly suggest a connection between this factor and the negative results obtained with the use of anti-varroa bottom boards during the 2000 trials. We can legitimately assume that the brood cluster temperature was lowered with the use of the opened anti-varroa bottom board. Numerous references can be found in scientific literature confirming that lower temperature conditions enhance the development of varroa populations. Ingemar Fries (12) states: "(...) mite population seems to grow faster in cooler climates than in warmer areas (...) it has been suggested that climatic factors are decisive in determining the mite population growth although the mechanism remains unclear." We can believe that a longer period of time in the capped brood stage resulting from a lower temperature favours an increase in the reproductive rate of the varroa mite's population. An increase of time in the capped brood stage enables the young female varroa mites to reach maturity before the bee emerges from its cell. Kraus and Velthuis (14) found that artificially reducing the brood temperature of colonies had the effect of doubling the mite population in comparison with control groups. Their laboratory tests allowed them to determine that 33 C was the optimal temperature for varroa mite reproduction. Kraus and Velthuis (14) suggest that beekeepers adopt practices that aid colonies in maintaining brood temperature at 35 C. The results obtained by Kraus and Velthuis were not available when planning for the 2000 trials as they were published in October of the same year. Reference to the influence of temperature on the rhythm of natural varroa drop can also be found in recent scientific literature. Thomas C. Webster (4) found that this drop is correlated to the average outdoor daytime temperature. J.T. Ambrose (13) also found (2001) that when infested adult bees were exposed to variable temperatures in laboratory conditions, the percentage of varroa mites falling from the bees increased with the elevation of the ambient temperature. Here again we can deduce that the brood chamber temperature should not be lowered.

It is already well known that Varroa are seen in greater numbers in strong colonies. This is counter intuitive, but may have its cause in the wintering of large colonies which have the ability to maintain brood through-out the year and for which the mites are profoundly grateful.

Perhaps a large colony in winter has enough bees to maintain a temperature high enough to maintain some brood, which not only sustains the varroa life cycle, but allows them to thrive with an increased pupation period, caused by the colder ambient temperature.

Experience with overwintered nukes being free of varroa is that they may have been cold enough for long enough to not be able to sustain any brood which disrupts the varroa reproduction cycle. This can only apply as long as the varroa cannot live for 5 months as an adult to carry it over a brood less period.

Very hot ambient temperatures are also disruptive to the varroa reproduction cycle, due to a shortened pupation period of bees.

A big colony that keeps brood through the winter assists the varroa reproduction cycle because there is brood, but the cold extends the pupation period. This allows more mites to mature and hatch which explains the increase in the numbers of varroa in such a colony.

The "27 days or many months" can therefore reasonably be queried. Perhaps they only live for 27 days under all conditions, but this could only be proven by taking out brood for a month in winter and seeing if varroa are still present at the end of it.

RESEARCH 3.4 INSECTICIDES

A.NEONICOTINOIDS.

The damage that this class of (kill all insects) pesticides appears to have done to the eco-system has become both profound and universal. There is no longer an uncontaminated base line to which damage can be compared. The only (dubious) benefit that we get is that we no longer get insect spattered car windscreens. The "crop protection" industry may be profitable, but it is a brutal and primitive method that will be superseded by genetically modified insect releases.

On page 11 of New Scientist (21 September 2013) an article written by Andy Coghan explains what must be the future. It describes the now proven method of genetically modifying the male insect of a pest species so that its female progeny die before it can reproduce. The male progeny continues to propagate the lethal gene through the population. It is simply bred and released into the environment.

Tony Nolan of Imperial College says that "It is strictly species-specific as the insects only seek out mates of their own species. Controlling the pests with insecticides would not be so discriminating".

B.MISTAKEN ASSUMPTIONS ON THE USEFULNESS OF HONEYBEES TO OIL SEED RAPE PRODUCTION

An agri-chemical witness (28/11/2012) stated (to the Environmental Audit Committee in the Houses of Parliament) that only about 10% of OSR is insect pollinated. Research in Pakistan (part copied below) indicates that it is considerably more at about 80% by yield.

Pak. Entomol. Vol. 31, No.2, 2009

THE POLLINATION BY HONEYBEE (APIS MELLIFERA L.) INCREASES YIELD OF CANOLA (BRASSICA NAPUS L.)

ABSTRACT

In order to quantify the response of honeybee on canola seed yield, an experiment was conducted at National Agriculture Research Center, Islamabad, during 2007-08, in complete randomized block design with two treatments (i. Plants caged with honeybees ii. Plants caged without honeybees) with four replications each. Number of pods set, pods weight, seed count in hundred pods and yield per plant were measured in twenty randomly harvested plants. The results showed significant increase in all the plant parameters caged with bees as compared to the plants without bees (control).

Number of pods and number of seed pods with pollination were 815 and 20 while without pollination was 349 and 15. The seed weight and yield (gm) with pollination was 26 and 7.6 (gm) and without pollination was 9.3 and 1.51 (gm), respectively. It is concluded from the experiment that honeybee visitation to the canola flowers is important for pollination and increasing seed yield.

Key words
honeybee, pollination, canola, seed yield.

To get the full picture of how important pollinator insects are to OSR the original research document (abstracted above) can be inspected where diagrams show unequivocal dependence upon them. See http://www.pakentomol. com/Downloads/Issues/2009-2/ent4-paper%20canola%20paperfinal[ms%20 munawar.pdf

Any beekeeper instinctively knows that 10% must be a gross under estimate and it seems to have been proven in Pakistan in 2009 where yield without honeybees would be only 20% of that with them. (i.e. 7.6 gms. compared to 1.51 gms.) The only obvious caution is that the density of honeybees in the cages may have been far higher than normally found, but this is an argument for more honey bees rather than less.

The chemical companies appear to have mismanaged farmer perceptions with great skill. Farmers have become chemical addicts in the mistaken belief that no insects are better than any, even pollinators, because they were only worth 10% of OSR tonnage. The reality is that they are worth about 80% and deserve better treatment, if only for enlightened self-interest. Our bees pollinate farmers OSR and in exchange (prima facie) they kill them.

If production tonnage of OSR were the sole criteria for good, it would appear that the chemical companies who produce neonicotinoids should stop doing so immediately and take up beekeeping.

A policy among farmers of mutual restraint in the use of these chemicals needs to be initiated. To act alone and not use them when others will do so is pointless. Only a change in the law will ensure the benefits of mutual restraint will accrue to all.

C. FLAWED METHODOLOGY FOR TESTING THE IMPACT OF NEONICOTINOIDS ON HONEYBEES.

There is a prima facie case to be made that the test methodology used is grossly underestimating the effects of neonicotinoids on honey bees and their larva.

Who designed the methodology for testing these chemicals on bees?

If you go to The Website "Beyond Pesticides" and watch the video "Killing Bees - Are Governments and Industry responsible" you will hear an American beekeeper describe the methodology used in the United States by their Environmental Protection Agency. It is the same as that used here in the U.K. and described in the EPPO 2010 document seen in "side effects on honey bees" on http://www.eppo. int/PPPRODUCTS/honeybees/honeybees.htm

The common testing regime is that 4 colonies of 10,000 bees each are put next to one Hectare of neonicotinoid treated OSR in the U.K. (or 2 and a 1/2 Acres in the U.S.A. The only difference is the metric/imperial units used). This cannot be chance. The suspicion is that they have been adopted by the regulatory authorities from the same author, but who was this and who accepted it?

A fullscale open field trial method has been used according to the EPPO 2010 guidelines. The study called "Side effects on honey bees" does not look for larval deaths at all. The field size can be as little as one hectare for 4 colonies of 10,000 bees each, which are required by the test methodology. This is unrealistic. That is 4 bees per square meter. It is worth about 5 minutes of forage per day. The bees will then go elsewhere for the rest of the day to forage on neonicotinoid free flowers. Bees forage over about 30 square kilometers (30 million square metres or 3,000 hectares) yet the regulatory test allows them only a quarter of a hectare of neonicotinoid laced OSR forage per colony.

The EPPO 2010 methodology for field tests was modified by "2013 Bees GD" by the EFSA page 213 "Design of a Field Study". Pages 215 and 216 indicate that the size of the open field study containing seed treated with Neonicotinoids can be 2 Ha. with 7 hives x 10,000 honey bees. For all practical purposes there is little difference between the two regimes. The bees, if not contained in a cage will range over 3000Ha. The burden of neonicotinoids on the bees will be (on average) 2/3000ths. of the forage available within the bees range. This is miniscule compared to reality which would require about 500Ha. rather than 2Ha.

A 2Ha. field will be foraged by the 7 hives x 10,000 honey bees for a very short time every day picking up very little neonicotinoid before foraging elsewhere.

As a scientific method it is nonsense.

The following approach needs to be adopted, being a caged small field trial. If the full field control area cannot be caged reduce the number of bees to an area that can be. The effect will be an accurate, scientific result.

OBJECTIVE

1. To determine the damage (if any) caused by a pesticide on honey bees.
2. To determine the oil seed rape yield from 3 different regimes.

METHOD

Erect 9# 50m long x 4.5m wide poly tunnels and cover them with bee proof netting. Cost approx. £4k.for each poly tunnel.

1. 3# to have oil seed rape in them whose seed has been dressed with neonicotinoid. When in flower insert mini nukes into each with a mated queen and a cup of bees.

2. 3# to have oil seed rape in them whose seed has not been dressed with neonicotinoid. When in flower insert mini nukes into each with a mated queen and a cup of bees. Deploy contact insecticides as required and at night only.

3. 3# to have oil seed rape in them whose seed that has not been dressed with neonicotinoid. When in flower insert mini nukes into each with a mated queen and a cup of bees.

OBSERVE

1. The death curves in each colony of adult bees and larva in each poly tunnel.

2. The oil seed rape yield in each poly tunnel by weight.

THE LOGICAL CONSTRUCT

A mini nuke is a microcosm of a full colony, but must be small enough to be self-sufficient in pollen and honey within a 50m.x 4.5m. poly tunnel. A poly tunnel 100m.long would be better to ensure survival from starvation. It could be sowed with 10% of plants other than OSR to ensure a balanced diet.

It must have enough pollen and honey forage in the poly tunnel to be self-sufficient as a closed system.

Neonicotinoids are apparently able to kill a target pest, the pollen beetle. If it does this why does it not kill other non-target pollen consuming insects such as pollinating bees? What is the difference between the insects biology that makes neonicotinoids fatal to the former, but not to the latter? When pollen beetle numbers (despite the seed dressing) reach 5 per plant it is recommended that an additional treatment with either a pyrethroid or a neonicotinoid at the green bud stage is given. This recommendation is expected soon to be dropped to 3 beetles per plant. Good for agri-chemical profits, but not for pollinators such as honey bees.

The existence of the DEFRA/CRD regulators by approving products, allow the agri-chemical companies to privatize the profits while socializing the damage they cause. The current regulatory regime is counter-productive. It seems not to protect the public or the eco-system. The companies must be happy with the current approval regime, because it acts as a firewall against civil action against them-selves.

A document prepared by the French Food Standards Safety Agency(AFFSA) for Thiamethoxam-based Cruiser 350 made by Syngenta Agro SAS.was carried out for Maize which bees do not visit, because it has no nectar. Maize is wind pollinated just as any other grass is. This was submitted to the AFSSA by Syngenta apparently without a "Commercial in Confidence" tag. Any similar study by Syngenta for OSR (which bees do visit) has its "Commercial in Confidence" tag still intact. Why there is a difference is puzzling, but there may be a sensible reason that is not apparent. That for maize showed no effect on bees.This is not a surprise as they do not visit it.

Page 15 of the AFSSA attachment also has the following text:

Laboratory tests. The laboratory data indicate that thiamethoxam and the metabolite CGA 322704 are highly toxic to adult bees

Thiamethoxam's toxicity for larval development was estimated using a laboratory test developed by INRA which is currently being validated.

The INRA data which is currently being validated when released in early 2013 will determine the "toxicity for larval development" and would seem to be important. It is surprising that it is not already known.

Information on the possible larval damage to honeybees by neonicotinoids appears not to be available yet must be a major point of possible damage being a neurotoxin.

From the RSPB website.

Is this talk about national bird declines just hot air?

No. As well as all these apparent disappearances of birds, there has been a serious, countrywide, decline in the numbers of many birds, including many well known and loved species such as the song thrush, skylark, lapwing and house sparrow.

This decline has been slow and gradual, rather than sudden. **Most of the declining species are farmland birds**. On the other hand, most woodland species such as the blue tit, nuthatch and great spotted woodpecker are still doing alright. However, declines may have started in woodland habitat also, with lesser spotted woodpecker and willow tit now red-listed because of their severe declines. Farmland birds… declining? Just a coincidence, surely.

CHAPTER 4

DESIGN AND CONSTRUCTION

You cannot better the world by talking to it.
Philosophy, to be effective, must be mechanically applied.
Buckminster Fuller

There are two fundamental parts to any hive, the external envelope and the internal frames. These are explained in order below, followed by the differences needed for use in the third world to where ZEST is entirely appropriate.

Refer to Drawing 1 for B.S. National width frames and Drawing 2 for B.S. Langstroth width frames.

A.DESIGN

External Envelope

See Drawings 1, 2, 6 and 7.

The ZEST beehive frames can be deployed in a horizontal ZEST hive or in various configurations of wood hive boxes, preferably converted to the ZEST hive design principles of:-

1. Top bee entry and trickle cross top ventilation. This avoids the cooling stack effect and allows just enough ventilation to prevent condensation.
2. An external hive envelope that reduces the colony burden of thermo-regulation by both insulating and providing a measure of thermal capacity (or thermal lag) to the hive envelope.
3. The drawing of natural comb on a frame lattice of bamboo, wood strips (with wax starter strips) or from plastic T-Bars where the tail of the T forms starter strips for the natural comb.

Wood hives have the benefit of being amenable to a split inspection for queen cells unlike the ZEST. If a queen cell inspection is carried out on a 100% frame basis it requires the inspection of every frame with brood on it, but only overcrowded hives swarm. If they become overcrowded they can do so very quickly.

An active ZEST management policy is called for which takes away brood, bees and honey to pre-empt a swarm and make controlled rather than uncontrolled increase.

The benefits of a ZEST are better tempered bees that are less likely to swarm and which do not die from disease in the winter. None have done so to date. The ZEST is a more comfortable hive for the bees to fulfil their needs, and without an excluder which encourages swarming. Honey can be constantly removed from a ZEST, which can be treated as a larder and which would block up the brood space. New or returned empty frames are put back in to keep the bees busy.

The paramount ambition of a ZEST beekeeper is to assist the bees in thermo regulating the brood nest in a warm dry envelope. Slugs are common in traditional hives, but none have been reported in a ZEST. The ZEST hive design facilitates warm and dry by having a thermal mass, insulation all round with top entry and trickle top ventilation which prevents any draft from the stack effect.

Conventional wood hives are visited by the beekeeper for the following reasons:
1. Spring clean, disease scan and to spread the brood.
2. Stimulation feed.
3. Weekly inspections to check for swarm cells, spread brood, to add boxes and move honey blocks into supers (if brood boxes are used as supers as well).
4. Assembling mostly capped honey to be taken by employing Porter bee escapes, enticing the bees down onto box of fresh comb of foundation.
5. Collecting honey two days later.
6. Returning wet frames to the hive for cleaning up after extraction.
7. Taking shook swarms, artificial swarms and nuclei.
8. Called to site to recover damage or swarms.
9. Feeding for winter and treating against Varroa.(which ZEST's have proven to be free of).

Some of these operations can be carried out at the same visit, but not on all of the hives at the same time. The logistics of actions on site visits can be compared to a complicated military campaign such as the D-day landings with a lot of travelling involved.

By comparison, the logistics of the ZEST hive management system is as shown in Drawings 8A and 8B. Apart from the compulsory visits expressed there the system can be "let alone". The insulated partition boards will need to be moved regularly to ensure a compact brood nest without crowding it. This is also the opportunity to take honey on a frame by frame basis when it is sealed, cutting it into tubs for removal, taking away, or to use the ZEST nucleus box with Canadian escapes to clear the bees. Preference based on experience has been to brush off bees and place the honey laden frames in the back of a car away from the bees. Take away and deal with it later.

Not only is the ZEST hive a third of the cost on a frame area match, the labour and time needed to gather it is also about a third.

Frames See Drawings 9 and 10.

There are three frame choices available determined by the context.

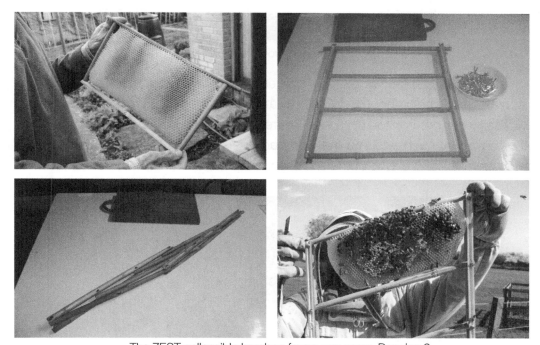

The ZEST collapsible bamboo frame sequence. Drawing 9.

The ZEST collapsible, bamboo, gravity frame has the merit of being suitable for making, using in and exporting from a third world country. It can be collapsed, shipped around the world, shaken out and put into ZEST's without further work. To make 200 of these frames by hand took about 12 hours. The pieces cost about 40p for each one, made in this country, but in China would be made from a free, in the garden, weed. A 2400 long cane makes one frame. A hive tool is not needed to lift it from the colony. The bees do not stick it down as much as the machined frame. Felt tip marks need to be made on the top frame bearer to space the frames.

The ZEST stapled wood frame. Drawing 10.

The ZEST stapled, wood, spacer, frames can be used if machine technology is available. It costs about 40p for the material and machining it. It was very strong vertically, but not horizontally when levered apart. Gimp pins were driven through the top bar at each end into the side bars to fix the joint laterally. 38 mm spacer nails are deployed on the same spacing principle as Hoffman frames.

The ZEST plastic frame.

The third and final choice preferred over all others is an injection moulded rigid PVC plastic frame. Plastic is a by-product of the fuel industry, but once obtained is recyclable. It is a useful legacy of the oil industry.

The first 2 frame types were tested in quantity during 2011. The plastic frame was tested during 2013 and 2014 proving the concept. Refinements of the prototype were initiated in 2014. They can be bought, but are not DIY. They can be purchased from www.thezesthive.com

The ZEST Hive mission commenced in late 2008 with the introduction of bamboo frames into a populated B.S.National brood box hive. The ambition was to develop a frame that could be deployed in the developing world which was made from locally available materials such as fishing line and bamboo. In a week wild comb had been drawn out and partially filled with blackberry honey.

From such a small and partial success (the fishing line stretched and sagged) locked-in preconceptions held for many years of what constituted acceptable beehive design and practice were shattered. A new design for a bee habitation was previously unperceived, but which was shockingly similar to that demanded for humans. They both need to be housed in a secure, warm, dry envelope where the young can be fed and raised at 35 degrees with an adequate supply of food and freedom from disease.

Existing beehive frame design and the enclosure to contain them did not meet these criteria. Such a realisation imposed a burden to speak and to design better.

The redesign of Beehives to allow for honeybee physiology

The ZEST frame can be deployed in a horizontal ZEST hive or in various combinations of traditional ones, preferably converted to the Zest Hive principles of:-

1. Top bee entry and trickle cross top ventilation. This avoids the cooling stack effect which is greater in winter.
2. An external hive envelope that reduces the colony burden of thermo-regulation by both insulating and providing a measure of thermal capacity (or lag) to the hive envelope.
3. The drawing of natural comb on a frame lattice of bamboo, wood strips or from plastic T-Bars where the tail of the T form starter strips for the natural comb.

It may be with or without a Durrant/ZEST insulated block encasement of standard wood hives.

Advantages of the ZEST plastic beehive frame

1. Half the cost, for the same comb area of a traditional wood frame.
2. A virtually forever frame.
3. No foundation needed. The bees act naturally to make their own comb. The cells tend to be smaller and the pupation time is reduced.
4. No assembly or winter maintenance needed.
5. Just unpack and drop into the hive. Hot wax can be added to the T-Bar tails.
6. Always hangs perfectly vertical.
7. Spacer tabs incorporated at top and bottom.

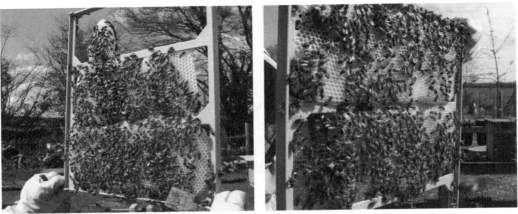

The Standard ZEST Beehive Plastic Frame (Grey)

The basic model.

The Delux ZEST Beehive Plastic Frame (White)

This is the same as the Standard frame, but has a food grade anti-microbial additive against bacteria and fungi added by SteriTouch. This is responsible for the higher price. We wish to determine by field testing if it does indeed supress EFB, Chalkbrood and Stonebrood.

We ask purchasers of the plastic ZEST frames to inform us on enquiries@thezesthive.com or Bill Summers on 01258 473015 if they find EFB, Chalkbrood or Stonebrood in the brood nests formed on white ZEST plastic frames.

Hives other than ZEST ones

The ZEST B.S. National width frame can also be inserted into:-
  a. Double B.S. National brood boxes.
  b. Triple B.S. National supers.
  c. A deep B.S. National with a super

The ZEST Beehive Plastic Frame Bearer Frame See Drawings 12A and 12B.

Plastic ones are available to purchase, being for deployment around the top edge of the ZEST hive to carry the frames.

Other sizes of frame

When the B.S. National width ZEST plastic frames are commercially proven, other sizes will be made available. These will be B.S. Langstroth width ZEST, National and Langstroth super frames.

The ZEST hive in use (shown with wood frames)

B.CONSTRUCTION

The original ambition of the ZEST hive design was one of being entirely DIY and for those with the skills it remains so, but this approach excludes the ZEST plastic frames which cannot be DIY and many beekeepers who do not wish to be. Do not be downhearted. Help is at hand. It will be exciting, starting here.

1. Always read the drawings first. Fully understand what is needed in both equipment and tools and have them all to hand. Night classes are not required.

2. Find some flat ground that is sunny and sheltered. Lay one 600x600 paving slab and level it after treating the ground with creosote to deter vermin from undermining it. You may need to level the slab again if it tips up in use. This is easily done with a long crowbar or similar device deployed as a lever. When the lever is deployed ram some earth under the slab with a spade at the low part until it is levelled. (I told you that this would be exciting).

3. Take 4# of the 440x215x100 heavy weight concrete blocks and lay them vertically (without mortar) in 2# x 2# rows so that the ends of the blocks cantilever equally out over the edges of the paving slab. Allow a gap of about 225 between the 2# rows. Orientate the long sides of the ZEST North/South so the sun gets on both sides at some time of the day.

4. Take 2# 1000x150 pieces of polythene such as that used as a damp proof course to guard against climbing vermin and lay it with equal overhang all round on the blocks. The only vermin so far found inhabiting a ZEST are ladybirds which fly in and spiders which abseil.

5. Take the 2# 1200x100x50 treated softwood floor support beams and lay them loose on the vermin guards. They will overhang equally at each end.

6. Take 6# of the 600/610/620x215x100 ultra lightweight insulating blocks by Durox, H+H Celcon, Tarmac or Ytong. Lay them flat on the floor support beams to form the floor.

7. Make and assemble the top bearer frame in wood or plastic, which supports the beehive frames and the roof blocks. Use it to ensure the accuracy of the void formed by the block walls by laying it on the floor and marking where the walls are.

8. Take 12# of the insulating blocks and build the walls. Stagger the joints vertically so that "bond" is achieved. There is no need to cut the end wall blocks which can simply hang out at the ends. (They will be useful to hang

the roof holding down ropes on when the roof is removed.) It is important that the void formed is 1140x374 for the B.S. National hive conversion and 1140x464 for the Langstroth hive conversion.

9. Lay the top frame bearer frame on the walls. Adjust the walls to fit the top bearer frame accurately.

10. Assemble the plywood partitions to match the hive cross section profile and ensure bee tightness.

11. Insert the ZEST beehive frames (If wood, then with wax starter strips. If plastic, then with melted wax brushed onto the tip of the T-Bar tails) and deploy two partitions as colony containment.

12. Deploy 25mm. foil backed insulation and deploy behind the partitions, away from the bees. Hold upright with a spare frame.

13. Take 6# of the 600/610/620x215x100 ultra lightweight insulating blocks (as the floor and walls) and lay evenly upon the top bearer frame. 2 of these will need a feed hole drilled in them which can be blocked up with a plastic wine cork or covered in a metal plate when not in use. When feeding deploy a tent peg in the hole to assist bee access to feed.

14. Purchase or find from a scrap yard, metal roof/wall profile section, cut to size and lay upon the block roof. There are about 200 of these to the ton as scrap which is worth about £200. Each metal roof is therefore worth £1. If you feeling rich or have someone in the household that demands good taste you can go for Duraline sheet which is 2000x950 and ideal for a ZEST roof,.....but almost as expensive as fire.

15. Hold down the roof sheet at each end of the ZEST with rope tied tightly under the floor beams. Where the rope passes over the 2 exposed metal sheet edges deploy a piece of hosepipe as protection. You can knot the rope to hold the protective hose in the correct positions.

16. Deploy the entrance blocks to suit the colony requirements. In winter shut it up almost entirely, leaving just enough to allow bee entry and trickle ventilation at one entry point.

17. Cut a further 2# insulating blocks at 374mm. for B.S.National and 464mm. for Langstroth to provide a winter partition between 2 colonies and seal it off at the top.

ZEST HIVE WITH WINTER PARTITION A pristine and final ZEST with its winter partition capped off at the top with wood to prevent colony mixing. A DIY wood top bearer frame is shown, but could be plastic.

The Durrant / ZEST hive by Dave Durrant

A transition stage in which traditional hives are deployed with ZEST frames within a ZEST envelope.

Under construction. See the box raising battens

Boxes and frames inserted.

Lid showing top entry and ventilation slots for winter mode.

Roof blocks added

Excluder shown on the summer top entry and ventilation frame.

Metal roof strapped down with summer top entry frame and with honey super above.

C.THE THIRD WORLD AND ZEST

A Declaration of Rights is, by reciprocity, a declaration of duties also.
Whatever is my right as a man is also the right of another; and it becomes my duty
to guarantee, as well as to possess.
Tom Paine

The sustainable principles and design of the ZEST hive remain sound for the Third World, which has a tendency to be in hot climates. The problem for the bees is less about keeping warm and more about keeping cool. Not only are insulated blocks not available in the Third World, using them would be detrimental to the bees wellbeing, whose need is to generally cool the colony, certainly in the day.

The answers to this dilemma are obvious. This design needs to be tested in the environment for which it is suggested. The changes listed below to the ZEST design for hot countries are a simplification of that for cold climates. Apart from these design changes, all else remains the same.

1. Use heavy weight concrete (or baked mud) blocks which are readily available in the Third World. These will tend to store the cool of the night and carry it through into the day. Conversely they will carry the heat of the day into the night. The extremes of temperature are moderated, alleviating the thermo-regulation demands upon the bees.

2. Incorporate an overhanging polythene sheet vermin guard onto a raised floor plinth to raise the walls on. Hot countries have termites and ants as pests to honeybees. A turned down plastic sheet at the floor/wall junction and covering the hive floor should act as a termite guard who cannot abseil round the edge of the plastic sheet.

3. Deploy wood load spreading battens around the top edge of the walls with bee access holes.

4. Use a raised metal "shade roof". This is a technique where the metal sheet roof is caused to "hover" above the frames with spacers allowing the void formed beneath it to be naturally ventilated. The sun falls on the metal sheet roof, heating it up, but this is dissipated by the ventilation under it, rather than be transmitted into the hive.

5. Use hand-made frames made from bamboo as in Drawing 9.

CHAPTER 5

MANUAL

If you have read this far you are either interested in honeybees and their welfare, a beginner beekeeper, a ZEST convert or someone with nothing else to read.

To the first I say you are wise to be concerned. There are a lot of beekeepers in a state of denial out there. To the second I say you are the ZEST missions unencumbered target audience. To the third I say I admire your decision to convert in the face of inertia and a natural desire to stay in bed and just hope that the ZEST beehive design analysis will all just go away. To the fourth I suggest joining a book club.

This book is directed at the 2nd and 3rd category. It is packed full of hard won practical, but exciting stuff such as creosoting the ground under the ZEST paving slab foundation to deter vermin using it as a roof to live under and consequentially undermining it. How much excitement do you want? There is more.

A.MANAGEMENT

Having built your ZEST it needs to have the bees installed as follows:

1. The ideal is to capture a swarm, simply tip it in and feed it. Conversion from traditional frames is then not an issue. Feed a swarm well with thick cane sugar syrup or a proprietary feed such as Apivert. The bees will draw out wax vertically with ease and enthusiasm, naturally working upside down. The bees have many workstation options, but will be centred on the cluster.

2. The second option is simply to put a colony on its traditional frames into your ZEST, let them settle in and build up, surrounded by ZEST frames onto which they will move. This requires more management and a little ruthlessness. The difficulties include the bees building unsupported wild comb on the bottom of the shorter traditional frames and an unwillingness to stop using the original frames. Early spring is the time to remove old frames that are not in use.

3. The final option is to carry out a shook swarm which has the merit of culling brood, comb diseases and varroa, but which sets back the strength of the

colony by culling the brood. I have not tried this, but it will work as well as it does in traditional hives.

Guidance on the use of Plastic Frames

The bees have been proven in the prototypes of this frame to be accepting and enthusiastic about drawing out their own comb. Brush hot wax onto just the tip of the tail of the T-bars to encourage the bees to draw the comb spine from them. There is a slight tendency for the bees to curve the comb around the centre of the brood rather than build it straight from end to end of the frame T-bar tail. Prevent this unruly behaviour and do so early. Press the comb back into place or remove it. The sooner this is done the better. If one frame is drawn askew the other adjoining ones will be drawn to match. Once you have a stock of well organised combs they will build further ones to match those.

You may already have traditional hives that are serviceable. They may even be newly bought. The ZEST principles of easing thermo-regulation, warm and dry, top entry and ventilation can still be deployed using them. Dave Durrant suggests that you encase your wood hives in insulating blocks and to make the hives top (of brood chamber) entry as suggested by Roy Pink with his Ventilated Integrated Bee Entry Zone (VIBEZ) as opposed to a BEASTE....Bottom Entry And STack Effect.

The frames can be deployed in a conventional double brood, deep brood and super or a 3 super configuration. This can be done with or without the ZEST principles of block encasement and top bee entry and ventilation,...but it is not advised.

THE VIBEZ IN ITS PURE FORM AND WITHOUT ENCASING BLOCKS

From below:-

Floor-Brood Box-VIBEZ-Queen Excluder-Honey Super-Cover Board-Roof
You can engage with these concepts to reduce your winter colony deaths while continuing to use your existing traditional hives within a thermo-regulating block envelope.

When the ZEST concept has been proven to you, you may wish to move on to the full ZEST hive design concept by removing the wood boxes and reconfiguring the blocks. The transition will be made easier by first deploying ZEST frames into any configuration of traditional wood boxes, ensuring the bees are settled before moving them into a full ZEST.

Insights have been gained and methods developed that can be passed on here. For example, bees locate their brood close to the entrance with the honey remote from it. If the entrance is at the top of the colony as it tends to be in a ZEST, the brood will be there as well. The traditional arch of honey and pollen over the brood is not the natural condition. With a top entry hive (as is normally found in a hollow tree) it is reversed. This may be because the colony can lose brood to predation and survive, but lose the honey stores and it will not. It is also warmer at the top, even with top ventilation and bee entry.

You cannot go far wrong if you trust the bee's natural instincts to resolve their own dilemmas. Their freedom to act in their own interests is as important as ours. The provision of wax foundation provides the bees with a pattern, but saves little on their need to produce wax. They do not draw it out so much as build on it. The cost/benefit of its provision is at best neutral. When bees draw wax out naturally from below they tend to make the worker comb cells smaller and reduce the egg to hatch period. This will impact on the varroa mite's reproductive efficacy. They need time in the pupa cell to mature. Interference by us in the bee's natural ways always has an impact and it is rarely beneficial… Except to feed them in famine and to cull or cure them when sick.

It soon becomes evident that ZEST colonies do not take up much time compared to traditional hives. ZEST's have plenty of room unrestricted by a queen excluder, being able to incrementally expand although they tend to build from the top down. They then move sideways in both directions and down. The queen has plenty of room to lay eggs and is unlikely to swarm. The (insulated) division boards are supplied and moved to accommodate the size of the colony. The aim is to make the colony compact, but not over crowded. The division boards are insulated on the side away from the colony which is especially useful in the winter to contain the colonies heat.

The day to day operational management of ZEST contained bees is clearly different to traditional hives which includes, but is not limited to the following:

1. Keep the colony contained sufficiently with the insulated partitions to allow the bees to warm up the hive, draw out wax and maintain the brood nest temperature with the assistance in a general way of the beekeeper adjusting the entrances for ventilation/cooling. They must not be allowed so much space and ventilation that they will have trouble thermo-regulating the brood nest. Just one entrance (or less) can be left open during the winter without getting condensation, the killer of bees.

2. Prevent swarming. ZEST hives are naturally slow to swarm. ZEST hives owned by the author has so far only made supersede queen cells except one which was a colony that the queen had accidently been taken away from to make a queen less nucleus for a queen cell nursery. The reason for swarming is normally hive congestion which does not occur in a ZEST with simple almost let alone management.

   There is a volume available in a ZEST that is the equivalent to 5 B.S. National brood chambers. It is best brought into use incrementally and in good time. A good honey flow in warm weather with a strong colony may bring a sudden need for up to 6 more frames. Be aware or suffer bad bee temper due to overcrowding. Maintain one (at least partially) empty frame at each end of the colony.

   The bees in a ZEST without partitions will tend to fill the top of the frames first and then work down. Putting in partitions will drive the bees down on the frames ensuring a more compact brood nest.

   The queen has insufficient room to lay eggs in a traditional hive restricted by a queen excluder and in which the brood nest becomes filled with pollen and honey. The ZEST has no such excluder and the bees can readily move the honey stores sideways or down to make way for brood. This has, so far, reduced the propensity to swarm to zero.

3. Maintain a regime of comb renewal. The tendency to reuse comb is strong with traditional frames due to the cost and effort of its replacement. The ZEST system is designed to be constantly replaced with natural comb. This is achieved throughout the year, but particularly at the end when the honey is taken and if the bees are fed for winter storage.

4. Ensure that sufficient stores are available throughout the winter. The method of assessing need in a wood hive is simple. Give the hive a lift at the back and experience will soon tell how much stores are in the hive. A ZEST is trickier to give a weight assessment lift, being heavy and

made of loose laid blocks, which tend to fall apart. Other methods and precautionary moves are needed.

The first is to harvest honey only from those frames that have no brood at all in them. This leaves honey in the hive which could have been taken, but was left for the bees. The ambition to not feed any artificial feed in ZEST's is possible by moving surplus frames of stores into ones with a deficit.

The second is to lift each frame slightly, assess the weights individually and roughly aggregate the total.

The third is to feed the ZEST's remotely and collectively with an inverted tub of cane sugar or Apivert bee feed in a plastic tub with straw in it to allow the bee's safe access.

The fourth is to feed the ZEST's individually with a contact feeder inverted over the feed holes in the roof blocks. Access for the bees from below to the gauze in the lid is greatly improved if hooked metal tent pegs or other plastic device is dropped into the feed holes. Smearing the gauze with honey assists its location as well.

The fifth is to start a stimulation feed around Christmas which serves as a precautionary feed if the stores are getting low. Repeat if/as it is used. In a traditional wood hive a plastic bag of baker's fondant with a smallish hole in it is placed over the feed hole in the cover board and the roof added. This seals the feed in with the bees. Such an arrangement with a ZEST will allow the feed to be open to the outside. A better arrangement for a ZEST is to fill honey jars with baker's fondant to contain it. Invert them over the feed holes in the roof blocks. A 12.5 Kilo box of it will fill 30 of them. The roof is then simply strapped down on top of the jar over the hole at one end and a 100mm concrete block laid flat at the other to support that end of the roof sheet.

Deploying fondant as winter/spring feed rather than liquid feed of any strength has the benefit of not leaking onto the blocks which tends to corrode them.

Having said all that ZEST hives seem to require less winter stores. This is not unreasonable as they can more easily be kept warm by the bees.

5. Accept the pressure from a ZEST to deploy "let alone" beekeeping methods. Visit them only to say hello, gloat, to show your friends and to adjust the insulated partitions to suit the colony size. Accept that much else is optional. A proper spring clean, maintenance and disease inspection is not. Manage further only upon apparent demand revealed by observation. Time normally spent beekeeping can be spent on the beach playing Frisbee and preparing for the Olympics. Carrying out a swarm cell inspection on a ZEST every 9 days is time consuming and not its best use. They strongly tend not to swarm so why do it.

6. Balance the colony strength by moving frames with sealed brood from strong colonies into weaker ones to supplement them, particularly in the spring. The bees are brushed off first of course. Overcrowding is prevented in the stronger colony and the weaker will make more honey. You will also need to take nukes from strong colonies so as to ensure an ordered succession and for sale or gift.

7. Improve the qualities of the queens. This is true of any sort of beekeeping and is important. ZEST colonies seem to be better tempered than others perhaps for the reasons that they are not spending nervous energy constantly thermo-regulating the brood nest. ZEST inspections do not take off the roof, break up the house beneath, chill the brood and then put

it all back together crushing the bees. The frames in a ZEST are removed one at a time. No bees need to be crushed at all. Proof of better temper is the non use of a smoker. It is habitually lit, but goes out from lack of need. A queen rearing program needs to be implemented. There are hundreds of books on the subject and none agree entirely with each other. The ZEST approach to breeding follows below.

8. Reduce bee deaths. For bee diseases see the book chapter "Objective (The Brief)" for the broad picture and to Neonicotinoids in the Research chapter. As a species we have a collective desire and interest in increasing colony and bee numbers in our own apiaries and across the planet. Our relationship with honeybees is symbiotic. While this cannot be denied it can be ignored by those who find it inconvenient to their business plan to wipe all insects from the planet.

At an operational level bee deaths can be avoided by:-
   a. Using only ZEST hives which are warm and dry rather than cold and damp.
   b. Making a frame by frame check for brood diseases in the spring clean and deal with it.
   c. Stimulation/emergency feeding after Christmas as already described above.
   d. Observe the bees. Are they taking in pollen? Are some weaker than the others?
   e. Deploy wasp traps from the spring onwards when the queens are foraging for themselves.

9. Make more bees and colonies. The ZEST hive is designed to contain two colonies overwintered in each by deploying the artificial swarm or nucleus method. It can be seen in Drawing 8A and 8B. It is intended that strong colonies be divided in June, July and early August to insure against winter loses. This gives options in the following spring not normally encountered. Deploying a ripe queen cell into the artificial swarm residue is possible, but adding a mated and laying queen is also an option. This allows the colony to establish itself more quickly and have the heart to fight off the attentions of wasps later in the year. The winter of 2013/2014 commenced with 21 ZEST's containing 27 colonies. 4 of the new colonies had had a ripe queen cell added. 3 of the new colonies had had a laying queen added.

3 of the 27 colonies were lost in winter from queen failure. 5 Nukes were also carried through the winter. None were lost. The summer of 2014 was needed not to make up lost numbers, but to deal with an embarrassing number of colonies in the spring.

10. Relax. It soon becomes evident that ZEST colonies do not take up much time compared to traditional hives. ZEST's have plenty of room unrestricted by a queen excluder. They are able to incrementally expand. They tend to build from the top down, starting where it is warmer. They then move sideways in both directions and down. They tend to store honey at each end of the brood nest and below it. Adjust to this natural self-management of the bees. The queens have plenty of room to lay eggs and are disinclined to swarm. The (insulated) division boards are easily moved to accommodate the size of the colony and to make the colony compact, but not over crowded. If it does get crowded the bees may not only think of swarming and get bad tempered, but may start to fill in the gap between the top of the frames and the underside of the roof blocks with brace comb. The powdering surface of most blocks usually prevents serious attachment. Pieces of 5mm plywood cover-board of the same size as the blocks will enable the ply to be "peeled" off of the frames without lifting the frames. The YTONG blocks from Germany are of a better, harder quality and are more readily attached to the frame tops.

    The division boards are insulated on the side away from the colony which is especially useful in the winter to contain the colonies heat.

11. Ensure that the roof sheet is strapped down well in a curve over the roof blocks and that the rope used is protected from fraying at the roof edges with short sections of hose pipe. Even if this is seemingly adequately carried out severe winds will require that one or both of the heavy concrete blocks made available for various uses on a ZEST are also laid on the roof sheets. Always deploy them throughout the winter and sleep soundly.

12. Make and deploy the Murchow rotator (Suggested by Stephen Murchow) to more easily inspect the frames and colour mark the queens. It is formed from a "Lazy Susan" more usually deployed in an Indian restaurant for dips

The Murchow rotator

Traditional wood hive management compares as being more intensive than for a ZEST, but retains some advantages.

Traditional wood hives need weekly split inspections looking for queen cells at the bottom of the frames in the top brood chamber. If none are seen frames of honey from the top brood box can be put above the excluder in the honey supers if they are also brood boxes. For flexibility one brood size boxes and frames can be deployed, but a strong back is needed. Removing brood frames that had clearly been designated by the bees for honey and replacing them with a new frame that the queen may get an opportunity to lay in allows brood box congestion to be reduced and with it the tendency to swarm. Replacing the frames also gives an opportunity to put in a drone/varroa trap used to cull varroa. This is the most effective technical method of reducing the varroa population in a traditional wood hive. It involves the insertion of a frame without any foundation in the top brood box at the third frame position in from one side. Presented with such an empty frame in a traditional hive in which all the foundation is worker, the bees draw down drone comb in the spring. This can then be cut out when sealed, culling the varroa contained within it. Varroa have 10-12 times greater tendency to lay eggs in drone cells than worker, because the drone has a longer pupation period, which allows the Varroa longer to mature. This varroa/drone trapping could be done several times in a year until the spring drone cell making season has passed. The additional merit of such a plan is that the bees do not litter drone comb so readily about the hive amongst worker comb on worker foundation. But it is all such a lot of hard work and requires a rigid timing discipline on a hive by hive basis.

The traditional method of extracting honey at a small scale from standard hives is also time and energy consuming by radial or tangential spinning of frames after uncapping, which preserves the frame and comb for reuse.

The disadvantages of this honey gathering method are:

1. It is messy and can lead to a divorce.
2. The frames when reused can transfer disease.
3. The cost of the extractor is high. The local bee keeping association's extractor is always busy being used by a member of the committee when you need it so the tendency is for everyone to buy one.
4. Unless the frames are carefully and evenly balanced the machine tries to leap about the kitchen, breaking the crockery.
5. Granulated honey is not harvested, but fed back to the bees.
6. Three site visits are needed, which are to install the Porter bee escapes, to take the honey and finally to return the wet frames to the bees.

The advantage of honey extraction from re-usable comb is that the bee's time and energy is saved by re-using the drawn comb. It is suggested that this allows them to collect more honey, putting it directly into already drawn out honeycomb. The savings in bee time claimed by deploying wax foundation however may be more apparent than real. There is more wax in honeycomb formed on foundation than that made naturally without it. They do not draw out foundation so much as use it as a template to draw it out on.

But this shoulder to the wheel and nose to the grindstone stuff can now all be history. ZEST is a better way.

The ZEST has no such management plan available, because:-
1. They cannot be split inspected.

2. They have not (so far) made queen cells that were not induced by blundering or queen failure. Taking off a queen nursery and accidently bringing away the queen did it nicely for me.
3. The ZEST bees can do what they want and need to in a free and unrestricted manner. This does not seem to include swarming. Why should it.
4. The ZEST is remarkably resilient to swarming being roomy and environmentally appropriate to the needs of the bees. They do supersede as they perceive the need
5. The ZEST hive has proven to be functionally varroa free.

The ZEST operating system has the merits of:-
1. The bees survive the winter with fewer stores.
2. Eliminating chemicals, pathogens and pests with the honeycomb removal.
3. Eliminates granulated honey (by melting) and pollen blocks blocking the queen laying eggs in the spring.
4. At season middle and end fresh worker comb will be built. This is ready for the queen to lay eggs in the following year after the food within them has been eaten.
5. Each ZEST can have 2 colonies through the winter giving normally unavailable options in the spring.
6. Leaves more honey in for the winter

The biggest test for the ZEST hive concept in practical use is the tidiness of the comb on the frames. The moveable frame concept is not negotiable, but requires a measure of discipline.

However twisted the comb is on plan, the bees will always defer to gravity in section. Gravity frames (that always hang vertically on a point support) allow the bee's natural inclination to build precisely vertical.

An appropriate ZEST frame design should cause the honeycomb to be:-
1. Drawn down parallel to each other at 34mm apart and not cross from one frame to the next. This is Hoffman spacing of 12 frames to a box.
2. Drawn down from the lowest point of the diameter of the bamboo frame.
3. Drawn down from the starter strips rather than an edge of the bar for the machined wood frames.
4. Drawn down off the plastic T-bars tails which are brushed with hot wax.
5. Not built anywhere other than in a moveable frame.

The ZEST was originally intended to be a hive of minimum provision with just one short central starter strip in a wood frame at three levels. The bees however tended to draw the line of the comb across the thickness of the frame horizontal bars after leaving the discipline of the (one) starter strip. Two strips as a minimum became the standard. Less provision was a saving too far. The bees were less than absolutely delighted to have their frames separated forcefully, but which was easily avoided with the use of 2 wax starter strips for each wood horizontal bar.

Bamboo and plastic do not need starter strips, because the bees will draw the spine of the comb from the lowest point on the horizontal bars of the frame.

ZEST hives do not suffer from varroa. During the 2012/13 season none were seen in any of the colonies and this was finally confirmed at end of 2014 when none could be found in the ZEST hive floor debris.

Conventional hives benefit from treatment with Thymol, because they have no natural resistance to Varroa infestation.

The ratio of Thymol crystals in grams compared to feed in grams is the starting point and can vary, according to Manley.

Manley suggests 1/12000 (Manley 1 strength) up to 1/3000 (Manley 4 strength). A maximum allowable dose beyond which damage to the bees may occur is not specified and may not be known. The author overdosed by a factor of 10 times and no harm to the bees seem to have occurred.

The lowest dose eliminates mould growth. The highest may kill varroa, but there does not appear to be any authoritative science on the subject.

The Thymol delivery system comprises dissolving the Thymol in surgical spirit at about 1 gram of Thymol to 4mls.of surgical spirit by shaking well in a bottle. This is the stock solution.

A dose of 1/12000 will require 5mls (a tea spoon) of the stock solution into a tub containing 12,000mls.or 12 litres.

A dose of 1/3000 will require 4# 5mls (4 tea spoons) of the stock solution into a tub containing 12,000mls.or 12 litres.

The overall management method is graphically described in Drawing 8A and 8B is as follows:-

1. From beginning of June until the end of August after the honey has been taken close off 6 of the 8 side bee entries and orientate the colony to one end of the ZEST. This is a prelude to taking off a nucleus or making an artificial swarm and a swarm residue that will be orientated at the other

end. In June, July make nuclei and early August make artificial swarms formed with the old queen on a frame of eggs in the colonies original position with the residue housed at the other end of the ZEST.

2. To the artificial swarm residue of brood and nurse bees positioned at the "other" end of the ZEST (with entry shut almost right down to prevent robbing) add a ripe bred queen cell for hatching and mating (or an already mini-nuke mated and laying queen, which has greater success and is quicker particularly for the years end). When the new queen is hatched and mated it will lay eggs in about 2 weeks if conditions are right.

3. Both the new and old colonies are heavily fed with Apivert after the honey has been taken at the end of July as usual.

4. The ZEST can then carry 2 colonies throughout the winter. The first with the old queen and the second with the new one. These are each housed on 6 or 7 ZEST frames at each end of the ZEST with a void between them formed by 2 insulated partitions (or a cut block winter partition capped off at the top). This gives a "no-man's land" between them which has vented access to outside, but not to each other.

5. In the spring the old queen can be removed and the two overwintered colonies united around the new queen. This gets off to a good, strong start and be ready to take advantage of an early spring flow. If one queen fails during the winter the two colonies can still be united around the surviving queen. Alternatively the old queen and its colony can be used to seed a new ZEST elsewhere.

The only attention the ZEST hive management system needs is to move the insulated division boards and the taking of surplus honey at the end of or during the year. It is important that the ZEST volume is sufficient to ensure that the colony is not overcrowded which will cause bad temper, but nor is it so roomy that the colony temperature cannot be maintained sufficiently to gain and maintain its critical mass. The volume in the winter should be reduced as far as possible and after the first frost just one access slot be allowed. If the correct volume in a ZEST is maintained in summer and the queen is young then there is little fear of swarming. "Let alone" bee keeping is then an appropriate management system. It is hard for an experienced beekeeper to shake off the tendency to manage weekly, which is often to the detriment of the bees and the harvest. ZEST beekeepers can spend more time on the beach. If in doubt just go away and leave it to the bees.

B.HARVESTING

Harvesting from a conventional hive tends to focus on honey, but ZEST beekeeping also includes harvesting wax, pollen and propolis. If this is all too much trouble the remaining mesh bag contents can simply be left out under cover for the bees to rob. Very little of anything will be left after 10 days.

There are 4 useful items that can be harvested. Honey, Wax, Pollen and Propolis. All are worth collecting separately from the ZEST hive which has aspirations to be the beehive of choice for citizens to engage with the post-industrial society of local self-sufficiency.

The ZEST way to take the frames is to work from both ends of the colony, brushing the bees off to carry them away, either directly indoors or into the back of a car. You can cut it out straight into a mesh bag in a honey tub with a lid and honey tap on it, but it is better to take it away first by leaning it in the boot of a car on a sheet. Leave any frame with brood with the bees.

A WOOD FRAME OF HONEY. Bees brushed off on site with a goose wing and carried away.

After uncapping the honey with an uncapping fork in a bee proof place so that all the cells are open, the comb is then cut out into a large course mesh bag in a 4 gallon wine bucket with a honey tap on it. Large areas of stored pollen can be cut out separately for making pollen pellets or crushed with the rest. This will add pollen to the honey which may or may not allow you to say it is" Honey with added pollen".

Take a clean stainless steel spade and seriously chop up the comb. Lift the bag out of the tub and hang it up above the tub. This will take 2 people. Let the bag drain for a couple of days giving it the odd lift and squeeze to turn it over a bit and get more honey running.

This will remove 90% of the honey which can be jarred.

The negative consequences of honey granulating in the hive becomes minimal, since it can all be melted at the end of the year in the ZEST honey warmer or at any other time,… or even in the following spring. 25 lbs. of granulated (blocking) ivy honey was extracted at the spring clean in 2011 from the first ZEST prototype. This was after 80 lbs. had been extracted in the previous autumn.

1. When the honeycomb is brought home from a ZEST it will have none of the disadvantages of conventional extraction. Extracting the honey from the comb is quick, easy and needs only a strong nylon mesh bag, plastic wine tub with a honey tap, an uncapping fork, a knife or spatula and a stainless steel spade. Take the honey frame to the tub, which has the nylon mesh bag in it. Stand the frame on its edge and uncap it on both sides with the fork. Cut out the panels of honeycomb with the knife into the nylon bag in the tub. Deploy the stainless steel garden spade to chop up the honeycomb. Lift and hang up the nylon bag on a meat hook above the tub and allow it to drain down. Leave for a day or so, alternately crushing and loosening the honeycomb in the bag. It runs out very quickly if broken up. If you find the honey is dripping off your elbows your technique needs enhancing, but with practice this will improve.

2. Separating the remaining honey, wax and dross can be done in the ZEST Honey Warmer. It has a heating element with a thermostatic controller under an open floor and can be obtained from:
Ecostat Unit 20 Beehive Workshops Parkengue Penryn Cornwall
TR10 9LX Tel: 01326 378654

3. It needs a bulk purchase of short nylon stockings, a two gallon stainless steel bucket and a 1 gallon saucepan with plenty of 10mm.dia.holes in

the bottom to fit inside it, forming a double saucepan. Alternatively a sieve can be inserted into the bucket into which the drained down comb filled stockings are then placed. The honey and wax runs more freely. Fill the stockings with as much compacted residue from the nylon draining bag as is possible. This may be assisted by pressing the residue into the stocking, stretched into a smaller saucepan. Tie a knot in the top of the nylons and place as many as possible inside the saucepan with holes or the sieve. Place that inside the 2 gallon saucepan and put both into the Honey Warmer. Using the thermostat, raise the temperature to honey melting temperature. Pour out the melted honey before raising the temperature to wax melting temperature. Pour the wax into moulds and discard the remaining dross in the nylon stockings.

One proud ZEST owner (Julie Challoner) spent some time considering and testing the absolute minimum harvesting technique that she could determine. It was for a piece of glass to be placed on a bowl which contained a sieve upon which the crushed honeycomb was placed. This was left out in the sun as a solar extractor. The bowl could be insulated with old clothing. The honey ran through the sieve, collecting in the bowl below. This is clearly suitable for owners of one ZEST hive and for third world use where resources are few.

Julie was also of the view that "surplus" honey taken in the autumn is not actually surplus until proven to be so in the spring when it should be taken as part of the spring clean. She also thought that taking pollen as the loaded bees enter the hive is unfair, but she had no such qualms of taking it from the honeycomb as pellets during its harvesting. See below.

RUNNING OUT THE HONEY

About 90% of the runny honey will run out of the nylon bags naturally. The remaining honey and wax can be removed by melting in the ZEST honey warmer. Wax to honey ration is about 1/30 by weight.

The majority of pollen can be removed from the honeycomb and processed separately as described below. It can also be broken up and left in the bag to drain through as "added" pollen… assuming the E.C. allows it of course.

DOUBLE SAUCEPAN

A stainless steel sieve of the right diameter can be deployed instead of the inner saucepan.

HONEY WARMER

Honey

The conventional use of honey is to eat it, praise the gods and raise your dental bills, but it has an external medicinal property as well. It can be used as a poultice that cures the flesh eating disease Necrotising Fasciitis, being a species of Streptococcus untouched by modern medicine. Is this hard to believe? It is easier if you have it. It was on the BBC so it must be true.

Wax

This is produced in abundance and floats on the last of the (heated) honey. When cooled and hardened it can be washed off and a candle maker found. There is currently little further use for bees wax, but if ZEST makes it more abundant then other uses may be found.

Pollen Pellets

The areas of stored pollen can be cut from the comb separately, laid on a couple of sheets of kitchen roll and placed in a microwave turned on for 2 minutes. The melted wax is soaked up by the kitchen roll and the pellets released to be collected. They roll around and do not stick together, perhaps due to the cell linings in which they were stored.

They can then be consumed as pellets (left) or turned back into pollen grains in a coffee grinder (right). It tastes bitter, but has a sweet background. Has a higher protein content than meat.

Propolis

This needs to be collected whenever the ZEST is opened. Just because it is free and not made by Big Pharma do not underestimate its value. It is well documented medicine that claims to boost the immune system and to prevent/cure/relieve arthritis and other maladies, morbidities and ailments. Drugs companies have investigated its possibilities. A small percentage part could not be analysed. As it was "discovered" rather than invented it could not be patented. They soon lost interest. Why wreck a billion pound income from arthritis drugs that are at best a palliative.

Let ZEST owners collect this bee product and dispense it to ailing friends and family with a daily prophylactic dose. It can do no harm and may do much good. Google propolis and see the research results. Much of it was in Russia where the research was publically funded and with no profit axe to grind.

C.QUEEN BREEDING

Queen rearing should be a part of any beekeeping management system.
Stock selection has been since Roman times for plants and animals and it has served us well. At a higher level it has become the field for experts and queen bee breeding is no exception. The best queens are selected and bred from in an isolated location where uncontrolled drones are prevented from entering. An island 10kms.from the mainland is a suitable place where standard tests for desirable qualities can be carried out, such as for temper, prolificacy, swarming tendency and honey production. Brother Adam started this and it continues, mostly abroad. This is a level of intellect and time commitment to which few of us can aspire, but we can buy from those who are inclined, and breed from them.

In such an isolated apiary both drones and queens can be selected. Those queens exhibiting the best characteristics are also encouraged to become drone layers, because it is through the drones that deterioration in temper qualities seem to occur.

Buying in breeding queens from professional breeding apiaries is effective, but breeding from them allows local uncontrolled drones to mate with the queen's progeny. This local drone mating provides new genetic material, but the most aggressive drones (as with all other animals) tend to win mating rights with the queen. This aggression seems to be carried into the workers at each successive generation. The first cross is quiet enough, the second tends to be less so and the third is aggressive.

The techniques best deployed to make queen cells are as follows:-

1. Put the breeder queen into a traditional wood hive with top entry and encased in insulated blocks as in the Durrant/ZEST shown on drawing 6. This has a greater winter survival prospect than a raw wood hive. It also has the advantage of the B.S. frames being easily manipulated with a Genter cage fitting into a traditional frame and allowing easy location of the queen.

2. Prepare the Genter cage as the instructions and fix it into a sheet of wax foundation in a brood frame.

3. Sunday. Open the round cap in the middle of the Genter cage face and lay the Genter cage frame down on the hive frame tops. Find the breeder queen, lay the frame on its back and scoop her up with a goose wing, holding her down upon it with your finger and place her through the hole in the Genter face. Keep the hole blocked until the queen is seen on the Genter cage comb face. Put the cap back on. She is caged. The workers will join her and commence drawing comb first time which can then be reused each time.

    (The cap can be replaced with a strip of aluminium bent twice so that it drops over the top of the Genter and holds the back and front together and also serves as a cap).

    Put the Genter cage frame back in the middle of the brood. Ensure that there is more space available in front of it than normal. If this is not done the bees may not be able to freely enter the Genter cage.

4. Monday. Check that eggs have been laid in the Genter cage. If yes, take the queen excluder Genter cage face off and allow the queen access to the rest of the brood chamber. If there is not a honey flow the bees may restrict the queen laying eggs. If so trickle feed them when the queen is in the Genter cage.

5. Wednesday. Go to a remote apiary (more than 3kms. from the breeder site) and find a strong colony to take a nursery from. Find the queen on a frame and set it aside so as not to inadvertently take her away. From the remaining frames find one with stores, four with hatching brood and one empty frame. Put these 6 frames (together with a couple more frames

of shook bees) into the ZEST nucleus travelling box and block the 2 top entries with sponge for travelling. Bring them back to the nursery site. When the bees have settled down remove the sponge so the bees can fly. Best results in the size and number of successful queen cells will be obtained in a congested nursery.

6. Thursday. Take the Genter cage frame out of the breeder queen hive with its recently hatched grubs and close up the frames and the hive. Shake or brush the bees off the Genter cage frame and take it to a warm car where the grubs can be transferred. Take each newly hatched grub (in its shiny, milky royal jelly) in its brown holder cup out of the back of the Genter cage and insert them into the brown plastic extension pieces. Insert the joined pieces into the yellow holder which hangs in the 2 test tube racks built into a standard brood frame. It will hold 2 rows of 10 cells. You will need up to 4 of these rack frames.

Take the empty frame out of the nucleus queen nursery and replace it in the middle of the nuke box with the 20 queen cells in the rack. Do not use spacers on this rack and press the other frames up against it to ensure crowding and warmth. (Take away the Genter cage frame parts and wash the remaining grubs out. When dry, reassemble and make ready for the next Sunday). Supply and keep topped up daily a honey jar feeder, with holes punched in the lid, over the feed hole while there are open queen cells in the nursery. The contents to be a sugar feed with added pollen or a substitute, shaken well before use. Better results in the size and number of successful queen cells will be obtained with this feeding.

7. 2nd Wednesday. Search the nursery for "wild" queen cells and cut them out. Failure to do so will result in one of them hatching earlier and killing the bred queen cells in the rack.

8. Sunday. Distribute queen cells to:-

a. Mini-nukes.
b. De-queened colonies.
c. Queen right colonies with an old queen,
for superseding. Protect cell.
d. Other beekeepers. Protect cell.

9. Leave one of the queen cells in the nursery after it has completed 2 sequential racks of queen cells. A two week cycle can be repeated every Sunday with two nurseries running at any one time and each nursery

receiving two sequential racks of queen cells. The second rack is often more successful, because more brood has hatched to form nurse bees supplying royal jelly. This gives a potential of 40 queen cells per week, but reality is half that number under even good conditions of high ambient temperature and honey flow.

Conditions that will be against good queen rearing and mating are as follows:-
1. Not buying in expensive queens from a professional breeder to breed from.
2. Only buying one queen and it failing, dying or being damaged. Buy two.
3. Too cold.
4. Too early in the year.
5. Not congesting the bees.
6. Not enough hatched drones which precede queen cell making
7. Putting too many frames with newly hatched grubs into a queen cell rearing nursery depleting the royal jelly available for the queen cells
8. The ambient temperature needs to be 22 deg.C for really good queen mating and with calm conditions.
9. Not spreading your mini-nukes randomly about the apiary. The queens returning to the wrong mini-nuke.
10. Forgetting to cut out the "wild" queen cells in the nursery colony.

The Queen cells can be either directly deployed into production hives or into mini-nukes for mating and laying before deploying into production hives as proven queens. This is not necessarily more successful than just putting in queen cells which the colony more readily adopts as their own.

In New Zealand where bee farmers have up to 10,000 hives the farmers may not see a queen all year. They simply make a lot of queen cells which are inserted into the colonies at the years end as supersede cells. They are protected from the bees with a short section of plastic conduit with a hacksaw cut down one end to allow the conduit to readily grip the (yellow) queen cell holder. The resident bees cannot reach the side of the cell where they would seek to break it down so their own queen can kill the new young queen in its cell. Once the young queen has hatched they can usually look after themselves, be mated and take over the hive in an orderly transition.

A regime of constant queen rearing at the Association level is a positive management technique that is a better than crisis management when queens fail.

The secret is to be ahead of events rather than events being ahead of you. Breed queens and use them. See the big picture. Play the numbers game.

CHAPTER 6

CONCLUSION

The ZEST Hive Project commenced as an exercise in minimum design for use in the Third World that does "MORE with LESS". The burning of fossil fuels, upon which the first world's standard of living is predicated, has never established itself securely in the Third World.

We are now standing on the brink of the biggest discontinuity in human history in the developed world as fossil energy resources run out. Civilisation may never recover. There can be no second industrial revolution based on fossil fuels.

Those who see the future will ask it not to come. We think that we will not suffer hardship, but our current wealth, health and ease in the First World is based on the burning of fossil fuels. In that respect the Third World may be a glimpse of the future and lessons in extreme frugality can be predicted.

Those who say that something will turn up to replace fossil fuel are wrong. To those who say that Governments will act, consider the record. To those who ask "Can a small group of dedicated individuals make a difference" G.K. Galbraith, when asked, replied that "It is the only way to do so".

Buckminster Fuller said that it will be done by design/science or not at all. And it will be done by making sense, not money.

Can we have a First World standard of living with Third World resources? Only with engineering ingenuity using renewable energy can it be so.

It is self-evident that food and energy security needs to be improved for a future when the benefits of fossil fuel become beyond the reach of ordinary people. It will lead to civil strife, for which we are poorly prepared. Security in food and energy can be ensured for the next generation and beyond by acting now to ensure the self sufficiency of small sustainable communities. This can be achieved while we still have the fossil fuel necessary to address the effects of its depletion. Our deposit (fossil) energy account can kick start our current (renewable) energy account. It will happen by driving events rather than be driven by them.

There have been political/social movements in the past such as The Diggers, The Levellers and The Tolpuddle Martyrs. They were started by some harsh reality. Each movement at the time addressed the issue of the day.

Today's issue of fossil fuel energy demise needs General System Engineers to design the machinery inspired by a descending "deposit" energy account and to switch to an ascending "current" energy account. Today's issue is to engineer a future not based on fossil fuels, but on their demise. This means social change that does not see the return of slavery, but a guarantee of freedom by working. It will mean a return to active citizenship rather than a dependency culture. Just saying that it is so will not make it so.

A bee colony is an example of a renewable energy system. Honey has energy that can be measured in calorific value. Unlike the compact energy already in fossil fuels this energy is widely distributed in the flowers of plants and trees, which rely on the sun. This dispersed energy is made compact when it is assembled into the hive by the bees. More honey energy is gained by the bees than is expended in gaining it. While the sun continues to shine we will have this renewable energy source.

Kahlil Gibran in his book "The Prophet" said it differently

Go to your fields and gardens and you shall learn that it is the pleasure of the bee to gather honey of the flower,

But it is also the pleasure of the flower to yield its honey to the bee.

For to the bee a flower is a fountain of life,

And to the flower a bee is a messenger of love.

Fossil fuel is a hard act to follow and its presence fleeting.

Slavery may return. The success of Wilberforce's campaign in Parliament against slavery was thought to be due to religious commitment and of the final triumphant goodness of man, but it was the industrial revolution that freed the slaves. Slavery was no longer economical in the face of steam power. It was a battle easily won.

With fossil fuel, globalisation prevails, with its enslavement of the non-unionised masses in a competitive downward global market. The prospect of "No free trade without free unions" is remote between sovereign states in a global market. If democracy ever looked like changing this it will be made unlawful.

Without fossil fuel, self-sufficiency and real democracy will have to replace it. If it does not it will be the AK 47.

The ZEST hive is a general systems design for a democratic future in a post-industrial society. Anyone can have one. It is cheap, appropriate and amenable to a more self-sufficient way of life. It is a living sustainable system, not a product. No one owns it. It is free. Take it. Use it. Have fun and remember that:

There is a tide in the affairs of men
Which, taken at the flood, leads on to fortune.
Omitted, all the voyage of their life
Is bound in shallows and in miseries,
On such a full sea are we now afloat,
And we must take the current when it serves
Or lose our ventures
William Shakespeare

CHAPTER 7

FREQUENTLY ASKED QUESTIONS

1. **What is the purchase cost of a ZEST hive compared to the traditional one?**

    The capital cost for providing a unit area of honeycomb in a ZEST is about a third of that for traditional hives. The maintenance and running costs of the ZEST hive is also a small fraction of the traditional.

2. **Is the ZEST a more labour intensive method for keeping bees?**

    No. It is less so. It can be a let-alone system except to just move the partition boards to give enough, but not too much space for the colony.

3. **Without foundation, is the comb built in a disorganised way?**

    The bees have been proven in the prototypes of this frame to be accepting and enthusiastic about drawing out their own comb. Brush hot wax onto just the tip of the tail of the T-bars to encourage the bees to draw the honeycomb spine from them.

    If they do decide to be unruly in comb building press the comb back into place or remove it. The sooner this is done the better. If the first frame is drawn askew the other adjoining ones will be drawn to match. Once you have a stock of well organised combs they will build further ones to match those.

4. **Does the need to rebuild the comb every time honeycomb is taken mean that there is less honey to take overall?**

    Yes. About 3% by weight of the crushed, squeezed and run out ZEST honeycomb is wax. About 1% of traditionally extracted honey is wax. You have more wax to sell rather than foundation to buy.

5. **Do you get many swarms?**

    None in 2013 or 2014. The queens were mostly in their first year. The ZEST has the equivalent space of 5 B.S. National brood box space. Many blue marked queens were superseded in 2014, but there were no known swarms from the authors ZEST's nor any from any known to him.

6.   Can the queen and brood be separated from the honey in a ZEST with queen excluders?

This was tried with 2 vertical plastic excluders, but the bees sealed them up. The bees naturally keep the brood together so an excluder is perhaps an incumbency. The ambition to formally separate honey and brood remains a work in progress, but is not seen as critical as it may encourage swarming.

The pollen can be separated from the hone and harvested as a food supplement. See the chapter on harvesting.

7.   Without being able to heft the ZEST how do you estimate how much stores it has for winter use?

By lifting each frame slightly, estimating the weight and counting it up. ZEST hives use much less stores in winter.

8.   Do the bees take the feed from an inverted contact feeder on the block roof when it is about 125mm above the top of the frames?

They can be slow to do so. It is greatly improved by dropping a bent aluminium tent peg down the feed hole and between the frames. This allows direct access for the bees to the underside of the feeder gauze. It can be further improved by putting honey on the gauze which attracts the bees by olfaction.

9.   How many of your ZEST colonies bees died in the winter of 2012/2013?

None out of 21 colonies, but 3 had failed queens.

10.   How many ZEST colonies died in the winter of 2013/2014?

None out of 27 colonies in 20 ZEST (7 doubles) and 6 in ZEST Nuke boxes, but 3 ZEST had failed queens.

11.   How do existing beekeepers see the ZEST?

There is no neutral ground or mixed feelings. Love it or hate it.

12.   Who likes the ZEST?

Beginners, Ladies, Engineers, Designers, the Poor, the Practical and those who understand the design principles of Buckminster Fuller who said that human progress is made by DOING MORE WITH LESS by deploying DESIGN/SCIENCE within a GENERAL SYSTEMS THEORY framework.

13. **Why is it so unattractive?**

Beauty is in the eye of the beholder, rather than in the object itself, which is neutral. It is certainly different and adopts the IKEA approach to design which is one of cheap, functional utility. It will become a classic in time like other pure utility objects have done.

14. **Does the ZEST affect varroa.**

It is free of it. Reduced pupation period caused by smaller natural cell sizes and a warmer more consistent environment are probably the cause.

CHAPTER 8

TECHNICAL DRAWINGS

The drawings here will forever remain a work in progress. They are merely a statement at a moment in time when it was thought that the simplest way of achieving maximum beneficial effect is that shown in the drawings.

A new material, management insight or technical failure will change everything. When this occurs it is necessary to step back and consider the whole and not just the part, because the whole is more than the sum of the parts. Those who see the benefit of taking on the ZEST mission and methods will see better ways.

If they do not the ZEST exercise has not entirely succeeded.

THE "ZEST" HIVE FOR B.S. NATIONAL CONVERSION

24 PLASTIC
ZEST FRAMES
(RECOMMENDED).

———————— D.I.Y BLOCKS
———————— D.I.Y WOOD
———————— D.I.Y INSULATION
———————— PLASTIC

COMPONENTS

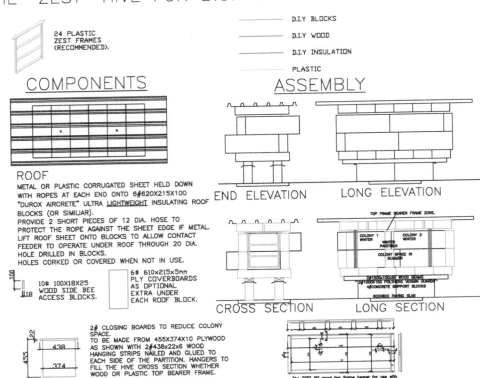

ASSEMBLY

END ELEVATION   LONG ELEVATION

ROOF

METAL OR PLASTIC CORRUGATED SHEET HELD DOWN
WITH ROPES AT EACH END ONTO 6#620X215X100
"DUROX AIRCRETE" ULTRA LIGHTWEIGHT INSULATING ROOF
BLOCKS (OR SIMILIAR).
PROVIDE 2 SHORT PIECES OF 12 DIA. HOSE TO
PROTECT THE ROPE AGAINST THE SHEET EDGE IF METAL.
LIFT ROOF SHEET ONTO BLOCKS TO ALLOW CONTACT
FEEDER TO OPERATE UNDER ROOF THROUGH 20 DIA.
HOLE DRILLED IN BLOCKS.
HOLES CORKED OR COVERED WHEN NOT IN USE.

10# 100X18X25
WOOD SIDE BEE
ACCESS BLOCKS.

6# 610x215x5mm
PLY COVERBOARDS
AS OPTIONAL
EXTRA UNDER
EACH ROOF BLOCK.

CROSS SECTION   LONG SECTION

2# CLOSING BOARDS TO REDUCE COLONY
SPACE.
TO BE MADE FROM 455X374X10 PLYWOOD
AS SHOWN WITH 2#438x22x6 WOOD
HANGING STRIPS NAILED AND GLUED TO
EACH SIDE OF THE PARTITION. HANGERS TO
FILL THE HIVE CROSS SECTION WHETHER
WOOD OR PLASTIC TOP BEARER FRAME.

The ZEST DIY wood top frame bearer for use with
B.S.National width ZEST's. Plastic also available

2# 20MM. PARTING BOARD INSULATION

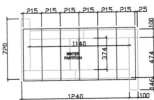

BLOCK WALLS AND FLOOR

FLOOR AND WALLS FORMED FROM 18#620X215X100
"DUROX AIRCRETE" ULTRA LIGHTWEIGHT INSULATING
BLOCKS. (OR SIMILIAR).
MOVEABLE INTERNAL WINTER PARTITION TO BE
2#374X215X100 CUT BLOCKS TO DIVIDE 2 COLONIES
OVER WINTER.
WOOD TOP CLOSER TO PARTITION TO BE AS BELOW
FOR WOOD TOP BEARER FRAME.
FOR PLASTIC TOP BEARER FRAME THE 12 DIMENSION
TO BE 3MM.

Guidance on building your ZEST hive
1. Clear the ground and level it. Treat it with something unpleasant, but not dangerous to prevent burrowing rodents from undermining the foundation slab.
2. Lay the 600x600x38 foundation paving slab. Level with a spirit level. (There is only one slab so that any uneven settlement can be easily dealt with by levering it up and repacking with earth to re-level it.)
3. Lay the 4# 440x215x100 heavy foundation blocks on the slab as shown on the drawing, being corbelled out from the foundation paving slab.
4. Lay 2#1000x150 polythene vermin guards on the foundation blocks. (The only vermin so far recorded in a ZEST are ladybirds, which fly in, and spiders which absell.)
5. Lay 2#1200x100x50 treated softwood as floor support beams.
6. Lay 6#620x215x100 ultra lightweight "Durox Aircrete" blocks on the beams to form the floor (or similiar other makes) These can also be used if lightweight, as well as other thicknesses of block.
7. Assemble the plastic or wood top frame bearer frame, lay it on the floor so that is equal in both directions and mark the internal volume of the ZEST hive on the floor.
8. Lay 12#620x215x100 blocks as the floor, but for the walls, as shown on the drawing and ensure that the vertical joints are staggered to achieve bond. No mortar.
(It is not neccessary to cut the end blocks to size. They can be left protruding out and used to hang the roof sheet holding down ropes when being removed.) The internal ZEST void dimensions on plan must be maintained to accept the top frame bearer.
9. Place the top frame bearer frame onto the top course of wall blocks.
10. Assemble and insert the plastic frames into the hive ensuring that they swing free.
11. Insert the closing boards where required to ensure sufficient space for the bees, but not too much. (It is important to make the closing boards accurately to ensure that the hive cross section profile is filled and is bee tight.)
12. Deploy the closing board insulation on the side away from the bees.
13. Place 6#620x215x100 roof blocks (as the floor) on the top frame bearer frame.Two of these blocks to have 20 dia. holes drilled in them to permit feeding from above. Insert a tent peg to assist bee access to the inverted feed.
14. Lay the roof sheet on the roof blocks. This can be scrap metal or a corrugated sheet.
15. Hold the roof sheet down with holding down ropes at each end with the rope passing under the floor beams. If a metal roof is used protect the rope at the contact points with plastic hose. When feeding, the roof sheet can be raised onto 2 blocks to allow clearance for the feeder.
16. Deploy the side and end bee access blocks for the conditions at the time. Winter requires just one side access open. Hot summer requires them all open.
17. Cut accurately and insert the moveable cut winter partition blocks to implement the ZEST Management plan shown on Drawing 8. Cap it off with wood blocking as drawn.

THE "ZEST" HIVE FOR B.S. LANGSTROTH CONVERSION

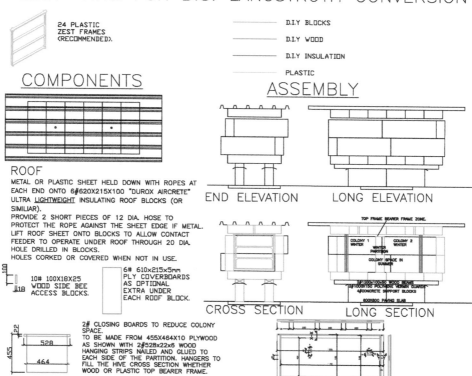

24 PLASTIC
ZEST FRAMES
(RECOMMENDED).

——————— D.I.Y BLOCKS

——————— D.I.Y WOOD

——————— D.I.Y INSULATION

——————— PLASTIC

COMPONENTS

ASSEMBLY

END ELEVATION   LONG ELEVATION

CROSS SECTION   LONG SECTION

ROOF

METAL OR PLASTIC SHEET HELD DOWN WITH ROPES AT EACH END ONTO 6#620X215X100 "DUROX AIRCRETE" ULTRA <u>LIGHTWEIGHT</u> INSULATING ROOF BLOCKS (OR SIMILIAR).
PROVIDE 2 SHORT PIECES OF 12 DIA. HOSE TO PROTECT THE ROPE AGAINST THE SHEET EDGE IF METAL.
LIFT ROOF SHEET ONTO BLOCKS TO ALLOW CONTACT FEEDER TO OPERATE UNDER ROOF THROUGH 20 DIA. HOLE DRILLED IN BLOCKS.
HOLES CORKED OR COVERED WHEN NOT IN USE.

10# 100X18X25 WOOD SIDE BEE ACCESS BLOCKS.

6# 610x215x5mm PLY COVERBOARDS AS OPTIONAL EXTRA UNDER EACH ROOF BLOCK.

2# CLOSING BOARDS TO REDUCE COLONY SPACE.
TO BE MADE FROM 455X464X10 PLYWOOD AS SHOWN WITH 2#528x22x6 WOOD HANGING STRIPS NAILED AND GLUED TO EACH SIDE OF THE PARTITION. HANGERS TO FILL THE HIVE CROSS SECTION WHETHER WOOD OR PLASTIC TOP BEARER FRAME.

2# 20MM. PARTING BOARD INSULATION

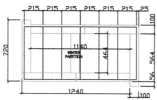

BLOCK WALLS AND FLOOR

FLOOR AND WALLS FORMED FROM 18#620X215X100 "DUROX AIRCRETE" ULTRA <u>LIGHTWEIGHT</u> INSULATING BLOCKS. (OR SIMILIAR)
MOVEABLE INTERNAL WINTER PARTITION TO BE 2#464X215X100 CUT BLOCKS TO DIVIDE 2 COLONIES OVER WINTER.
WOOD TOP CLOSER TO PARTITION TO BE AS BELOW FOR WOOD TOP BEARER FRAME.
FOR PLASTIC TOP BEARER FRAME THE 12 DIMENSION TO BE 3MM.

The ZEST DIY wood top frame bearer for use with B.S.Langstroth width ZEST's. Plastic also available

Guidance on building your ZEST hive
1. Clear the ground and level it. Treat it with something unpleasant, but not dangerous to prevent burrowing rodents from undermining the foundation slab.
2. Lay the 600x600x38 foundation paving slab. Level with a spirit level. (There is only one slab so that any uneven settlement can be easily dealt with by levering it up and repacking with earth to re-level it.)
3. Lay the 4# 440x215x100 heavy foundation blocks on the slab as shown on the drawing, being corbelled out from the foundation paving slab.
4. Lay 2#1000x150 polythene vermin guards on the foundation blocks. (The only vermin so far recorded in a ZEST are ladybirds, which fly in, and spiders which abseil.)
5. Lay 2#1200x100x50 treated softwood as floor support beams.
6. Lay 6#620x215x100 ultra lightweight "Durox Aircrete" blocks on the beams to form the floor (or similiar other makes) These can also be used if lightweight, as well as other thicknesses of block.
7. Assemble the plastic or wood top frame bearer frame, lay it on the floor so that is equal in both directions and mark the internal volume of the ZEST hive on the floor.
8. Lay 12#620x215x100 blocks as the floor, but for the walls, as shown on the drawing and ensure that the vertical joints are staggered to achieve bond. No mortar.
(It is not necessary to cut the end blocks to size. They can be left protruding out and used to hang the roof sheet holding down ropes when being removed.) The internal ZEST void dimensions on plan must be maintained to accept the top frame bearer.
9. Place the top frame bearer frame onto the top course of wall blocks.
10. Assemble and insert the plastic frames into the hive ensuring that they swing free.
11. Insert the closing boards where required to ensure sufficient space for the bees, but not too much. (It is important to make the closing boards accurately to ensure that the hive cross section profile is filled and is bee tight.)
12. Deploy the closing board insulation on the side away from the bees.
13. Place 6#620x215x100 roof blocks (as the floor) on the top frame bearer frame. Two of these blocks to have 20 dia. holes drilled in them to permit feeding from above. Insert a tent peg to assist bee access to the inverted feed.
14. Lay the roof sheet on the roof blocks. This can be scrap metal or a corrugated sheet.
15. Hold the roof sheet down with holding down ropes at each end with the rope passing under the floor beams. If a metal roof is used protect the rope at the contact points with plastic hose. When feeding, the roof sheet can be raised onto 2 blocks to allow clearance for the feeder.
16. Deploy the side and end bee access blocks for the conditions at the time. Winter requires just one side access open. Hot summer requires them all open.
17. Cut accurately and insert the moveable cut winter partition blocks to implement the ZEST Management plan shown on Drawing 8. Cap it off with wood as drawn.

2

THE "ZEST" NUKE BOX, HARVESTER AND CARRIER FOR B.S. NATIONAL

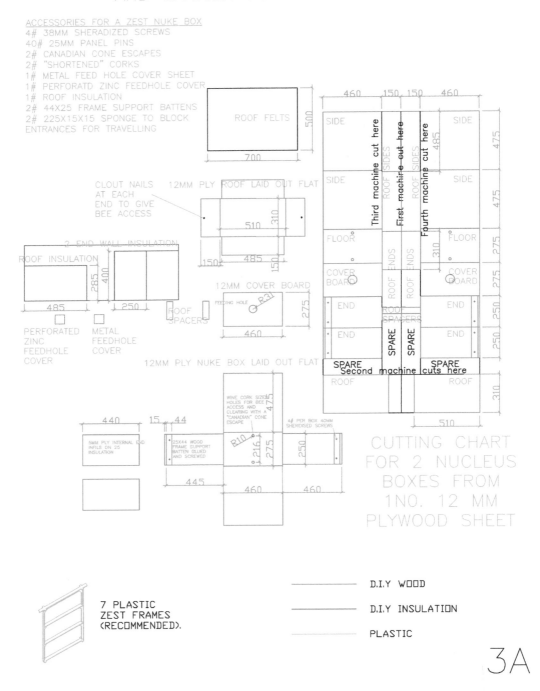

ACCESSORIES FOR A ZEST NUKE BOX
4# 38MM SHERADIZED SCREWS
40# 25MM PANEL PINS
2# CANADIAN CONE ESCAPES
2# "SHORTENED" CORKS
1# METAL FEED HOLE COVER SHEET
1# PERFORATD ZINC FEEDHOLE COVER
1# ROOF INSULATION
2# 44X25 FRAME SUPPORT BATTENS
2# 225X15X15 SPONGE TO BLOCK
ENTRANCES FOR TRAVELLING

ROOF FELTS

500

700

CLOUT NAILS   12MM PLY ROOF LAID OUT FLAT
AT EACH
END TO GIVE
BEE ACCESS

510   310

2 END WALL INSULATION

ROOF INSULATION

285   400

485   250

PERFORATED   METAL
ZINC   FEEDHOLE
FEEDHOLE   COVER
COVER

150   485   150

12MM COVER BOARD

ROOF
SPACERS

FEEDING HOLE   R31

460

275

12MM PLY NUKE BOX LAID OUT FLAT

440   15 44

5MM PLY INTERNAL END
INFILS ON 25
INSULATION

25X44 WOOD
FRAME SUPPORT
BATTEN GLUED
AND SCREWED

445

WINE CORK SIZE
HOLES FOR BEE
ACCESS AND
CLEARING WITH A
"CANADIAN" CONE
ESCAPE

R10   216   275   47

4# PER BOX 40MM
SHERIDISED SCREWS

250

460   460

460   150   150   460

SIDE   SIDE   475

Third machine cut here
First machine cut here
Fourth machine cut here

ROOF SIDES
ROOF SIDES
ROOF SIDES

SIDE   SIDE   475

485

FLOOR   FLOOR   275

ROOF ENDS
ROOF ENDS

310

COVER   COVER   275
BOARD   BOARD

END   END   250

ROOF
SPACERS

END   END   250

SPARE
SPARE

SPARE   SPARE
Second machine cuts here

ROOF   ROOF   310

510

CUTTING CHART
FOR 2 NUCLEUS
BOXES FROM
1NO. 12 MM
PLYWOOD SHEET

7 PLASTIC
ZEST FRAMES
(RECOMMENDED).

─────────── D.I.Y WOOD

─────────── D.I.Y INSULATION

─────────── PLASTIC

3A

The B.S. National ZEST Converted Poly Hives

Deploy the Durrant/ZEST double brood box using your existing equipment as on Drawing 6
OR
Paynes double B.S. Brood Polyhive adjusted to suit ZEST design principles.

1. Top bee entry and trickle cross top ventilation. This avoids the cooling stack effect.
2. An external hive envelope that reduces the bee's burden of thermo-regulation by both insulating and providing a measure of thermal capacity to the hive envelope.
3. The drawing on natural comb on a frame lattice of bamboo, wood strips or from plastic T-Bars where the tail of the T form starter strips for the natural comb.

Discard the floor, entrance block, metal runners. Amend and replace with the following.

1. 432x418x12 plywood floor on any base of insulation, concrete blocks or both.
2. Position first bottom brood box to notch over plywood floor.
3. Infill frame lug recesses between boxes with 418x33x24 wood to form flush surfaces inside.
4. Add second top brood box locking down onto wood infill pieces. Cut bee entry and ventilation slots on each of 4 sides in the top edge at $\frac{1}{4}$ from the right end. Ensure clear bee access under the roof. The entrys can then be opened or closed as required to receive trickle top ventilation as a full colony or as 2 nuclei. The slots are formed with a hacksaw blade and a sharp knife. See 8. below for divider.
5. Add strips of thin 418x33 Perspex to the frame bearing ledges to bear the frame lugs upon and spread the load.
6. Add the 12 ZEST plastic frames.
7. Add entrance blocks as required.
8. Insert a 434 wide x 448 high overall x 5 ply divider cut to match the hive inside profile and deployed as a partition between 2 nuclei. Form 2 # 16x33 ears at top to hang the partition on.
9. Add entrance blocks as required.
10. Add the Perspex cover.
11. Add the roof.
12. Add a brick on it.

Paynes double depth B.S. Polyhive Nucleus box with 2 Ekes adjusted to suit ZEST design principles.

1. Set Nucleus down on a base.
2. Insert a Perspex floor to cover the ventilated floor.
3. Add 2 ekes.
4. Notch top of eke to allow top bee entry and ventilation at both ends. They can be opened or closed off as required to receive trickle top ventilation The slots are formed with a hacksaw blade and a sharp knife.
5. Add strips of thin 215x33 Perspex to the frame bearing ledges to bear the frame lugs upon.
6. Add the 6 ZEST plastic frames.
7. Add the Perspex cover.
8. Add entrance blocks as required.
9. Add the roof.
10. Add a brick on it.

3B

THE "ZEST" NUKE BOX, HARVESTER AND CARRIER FOR B.S. LANGSTOTH

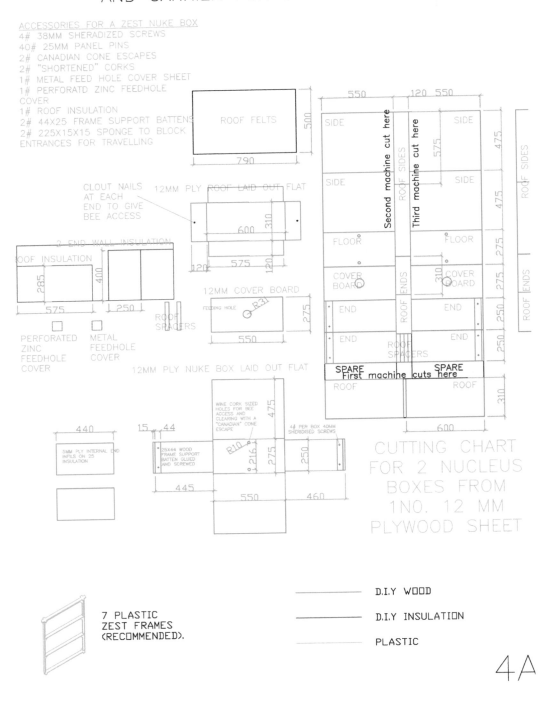

ACCESSORIES FOR A ZEST NUKE BOX
4# 38MM SHERADIZED SCREWS
40# 25MM PANEL PINS
2# CANADIAN CONE ESCAPES
2# "SHORTENED" CORKS
1# METAL FEED HOLE COVER SHEET
1# PERFORATD ZINC FEEDHOLE COVER
1# ROOF INSULATION
2# 44X25 FRAME SUPPORT BATTENS
2# 225X15X15 SPONGE TO BLOCK ENTRANCES FOR TRAVELLING

ROOF FELTS

CLOUT NAILS AT EACH END TO GIVE BEE ACCESS

12MM PLY ROOF LAID OUT FLAT

2 END WALL INSULATION
ROOF INSULATION

PERFORATED ZINC FEEDHOLE COVER

METAL FEEDHOLE COVER

ROOF SPACERS

12MM COVER BOARD

FEEDING HOLE

12MM PLY NUKE BOX LAID OUT FLAT

WINE CORK SIZED HOLES FOR BEE ACCESS AND CLEARING WITH A "CANADIAN" CONE ESCAPE

5MM PLY INTERNAL END INFILLS ON 25 INSULATION

25X44 WOOD FRAME SUPPORT BATTEN GLUED AND SCREWED

4# PER BOX 40MM SHERDISED SCREWS

SIDE — SIDE
ROOF SIDES
SIDE — SIDE
FLOOR — FLOOR
COVER BOARD — COVER BOARD
ROOF ENDS
END — END
END — END
SPARE — SPARE
First machine cuts here
ROOF — ROOF

Second machine cut here
Third machine cut here

ROOF SIDES
ROOF ENDS

CUTTING CHART FOR 2 NUCLEUS BOXES FROM 1NO. 12 MM PLYWOOD SHEET

7 PLASTIC ZEST FRAMES (RECOMMENDED).

D.I.Y WOOD

D.I.Y INSULATION

PLASTIC

4A

The B.S. Langstroth ZEST Converted Poly Hives

Deploy the Durrant/ZEST double brood box using your existing equipment as on Drawing 6

OR

Paynes double B.S. Brood Polyhive adjusted to suit ZEST design principles.

1. Top bee entry and trickle cross top ventilation. This avoids the cooling stack effect.
2. An external hive envelope that reduces the bee's burden of thermo-regulation by both insulating and providing a measure of thermal capacity to the hive envelope.
3. The drawing on natural comb on a frame lattice of bamboo, wood strips or from plastic T-Bars where the tail of the T form starter strips for the natural comb.

Discard the floor, entrance block, metal runners. Amend and replace with the following.

1. 522x418x12 plywood floor on any base of insulation, concrete blocks or both.
2. Position first bottom brood box to notch over plywood floor.
3. Infill frame lug recesses between boxes with 418x33x24 wood to form flush surfaces inside.
4. Add second top brood box locking down onto wood infill pieces. Cut bee entry and ventilation slots on each of 4 sides in the top edge at $\frac{1}{4}$ from the right end. Ensure clear bee access under the roof. The entrys can then be opened or closed as required to receive trickle top ventilation as a full colony or as 2 nuclei. The slots are formed with a hacksaw blade and a sharp knife. See 8. below for divider.
5. Add strips of thin 418x33 Perspex to the frame bearing ledges to bear the frame lugs upon and spread the load.
6. Add the 12 ZEST plastic frames.
7. Add entrance blocks as required.
8. Insert a 524 wide x 448 high overall x 5 ply divider cut to match the hive inside profile and deployed as a partition between 2 nuclei. Form 2 # 16x33 ears at top to hang the partition on.
9. Add entrance blocks as required.
10. Add the Perspex cover.
11. Add the roof.
12. Add a brick on it.

Paynes double depth B.S. Polyhive Nucleus box with 2 Ekes adjusted to suit ZEST design principles.

1. Set Nucleus down on a base.
2. Insert a Perspex floor to cover the ventilated floor.
3. Add 2 ekes.
4. Notch top of eke to allow top bee entry and ventilation at both ends. They can be opened or closed off as required to receive trickle top ventilation The slots are formed with a hacksaw blade and a sharp knife.
5. Add strips of thin 215x33 Perspex to the frame bearing ledges to bear the frame lugs upon.
6. Add the 6 ZEST plastic frames.
7. Add the Perspex cover.
8. Add entrance blocks as required.
9. Add the roof.
10. Add a brick on it.

4B

THE "ZEST" ELECTROSTATIC HONEY WARMER AND QUEEN CELL REARER

CUTTING CHART FOR 2 HONEY WARMERS FROM 1NO. 12MM PLYWOOD SHEET

Heating element can be purchased from "Ecostat" (01326 378654) with a thermostatic plug top. If greater accuracy is needed for queen rearing or accurate honey and wax melting obtain heating element wire from Ecostat and a digital thermostatic control unit.

2no. 310x44x32mm wood battens for supporting 4# 310x40x12 ply floor battens and forming a void for the heating element wire to be entirely suspended on silicon rubber bands supplied for that purpose

12no. 310x40x5mm ply battens for between jar layers

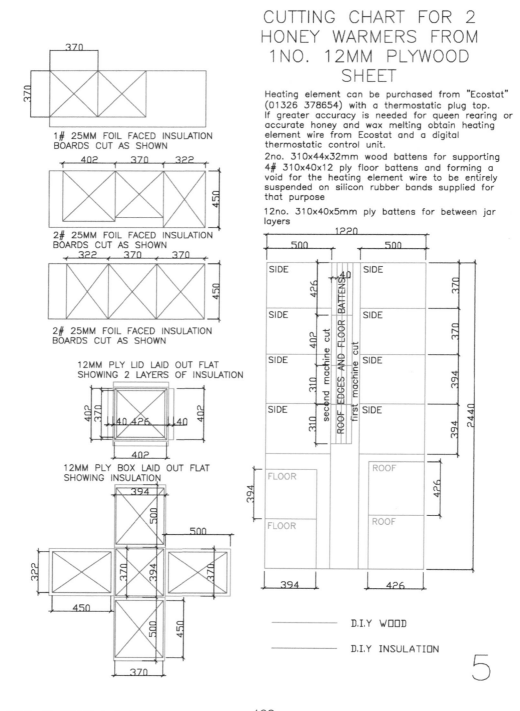

1# 25MM FOIL FACED INSULATION BOARDS CUT AS SHOWN

2# 25MM FOIL FACED INSULATION BOARDS CUT AS SHOWN

2# 25MM FOIL FACED INSULATION BOARDS CUT AS SHOWN

12MM PLY LID LAID OUT FLAT SHOWING 2 LAYERS OF INSULATION

12MM PLY BOX LAID OUT FLAT SHOWING INSULATION

D.I.Y WOOD

D.I.Y INSULATION

5

The ZEST/wood hybrid by Dave Durrant incorporating
The Ventilated Intermediate Bee Entry Zone (VIBEZ) by Roy pink

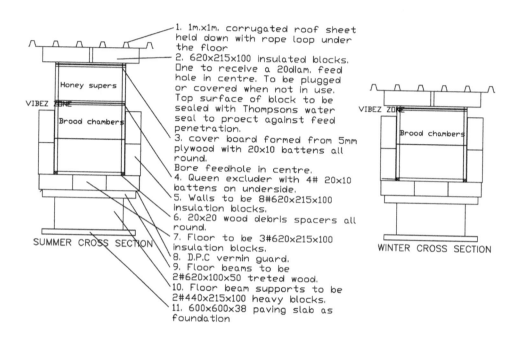

SUMMER CROSS SECTION

WINTER CROSS SECTION

1. 1m.x1m. corrugated roof sheet held down with rope loop under the floor
2. 620x215x100 insulated blocks. One to receive a 20diam. feed hole in centre. To be plugged or covered when not in use. Top surface of block to be sealed with Thompsons water seal to proect against feed penetration.
3. cover board formed from 5mm plywood with 20x10 battens all round. Bore feedhole in centre.
4. Queen excluder with 4# 20x10 battens on underside.
5. Walls to be 8#620x215x100 insulation blocks.
6. 20x20 wood debris spacers all round.
7. Floor to be 3#620x215x100 insulation blocks.
8. D.P.C vermin guard.
9. Floor beams to be 2#620x100x50 treted wood.
10. Floor beam supports to be 2#440x215x100 heavy blocks.
11. 600x600x38 paving slab as foundation

Ventilation and bee entries

4#395X20X10 wood battens glued and screwed to underside of Waldron excluder to form staggered entrances beneath.
4# entrance blocks.

EXCLUDER
VIBEZ SUMMER PLAN

Ventilation and bee entries

Feedhole

COVER BOARD
VIBEZ WINTER PLAN

6

THE "ZEST" HIVE BASIC DESIGN

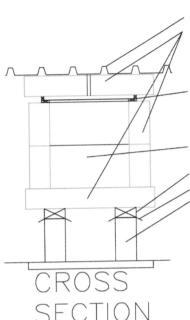

The Merits of a ZEST hive
1. Metal roof sheet from scrap.
2. Ultra light insulation wall blocks used as roof, floor and walls to insulate the colony and assist in thermo regulation.(Feedholes through roof blocks)
3. Top frame bearer frame with multiple top entrances for bee access and for adjustable trickle cross ventilation, housing up to four colonies. No stack effect to cool colony.
4. Void for one of three types of ZEST gravity frames. i.e wood, bamboo or plastic.
5. Wood bearers on an overhanging vermin guard / damp proof course.
6. Concrete blocks on a single paving slab raising the work position.

Notes:
   Maximum weight lift of 4kgs. honey in frame.
   Management to be 2 overwintered colonies. One with new queen and one with old.

CROSS SECTION

ALTERNATIVE FRAME TYPES
DIY STAPLED WOOD SPACER FRAMES FOR MACHINE TECHNOLOGY.
DIY COLLAPSIBLE BAMBOO GRAVITY FRAME FOR THIRD WORLD USE.
PLASTIC FRAME.

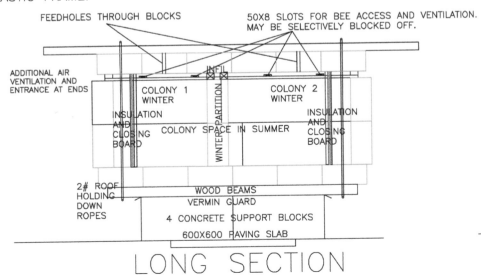

FEEDHOLES THROUGH BLOCKS

50X8 SLOTS FOR BEE ACCESS AND VENTILATION. MAY BE SELECTIVELY BLOCKED OFF.

ADDITIONAL AIR VENTILATION AND ENTRANCE AT ENDS

INFIL

COLONY 1 WINTER

COLONY 2 WINTER

INSULATION AND CLOSING BOARD

COLONY SPACE IN SUMMER

WINTER PARTITION

INSULATION AND CLOSING BOARD

2# ROOF HOLDING DOWN ROPES

WOOD BEAMS
VERMIN GUARD
4 CONCRETE SUPPORT BLOCKS
600X600 PAVING SLAB

LONG SECTION

7

THE "ZEST" ARTIFICIAL SWARM MANAGEMENT PLAN

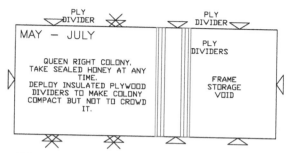

PLY
DIVIDER

PLY
DIVIDER

MAY — JULY

QUEEN RIGHT COLONY.
TAKE SEALED HONEY AT ANY
TIME.
DEPLOY INSULATED PLYWOOD
DIVIDERS TO MAKE COLONY
COMPACT BUT NOT TO CROWD
IT.

PLY
DIVIDERS

FRAME
STORAGE
VOID

FOR A WEEK BEFORE MAKING THE ARTIFICIAL SWARM ORIENT THE
COLONY ENTRY POINTS TO ONE END OF THE ZEST AS SHOWN

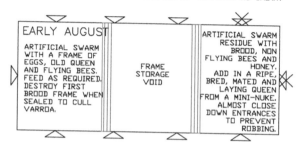

EARLY AUGUST

ARTIFICIAL SWARM
WITH A FRAME OF
EGGS, OLD QUEEN
AND FLYING BEES.
FEED AS REQUIRED.
DESTROY FIRST
BROOD FRAME WHEN
SEALED TO CULL
VARROA.

FRAME
STORAGE
VOID

ARTIFICIAL SWARM
RESIDUE WITH
BROOD, NON
FLYING BEES AND
HONEY.
ADD IN A RIPE,
BRED, MATED AND
LAYING QUEEN
FROM A MINI-NUKE.
ALMOST CLOSE
DOWN ENTRANCES
TO PREVENT
ROBBING.

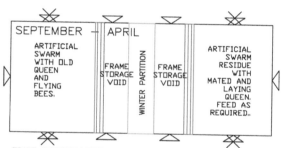

SEPTEMBER — APRIL

ARTIFICIAL
SWARM
WITH OLD
QUEEN
AND
FLYING
BEES.

FRAME
STORAGE
VOID

WINTER PARTITION

FRAME
STORAGE
VOID

ARTIFICIAL
SWARM
RESIDUE
WITH
MATED AND
LAYING
QUEEN.
FEED AS
REQUIRED..

FRAME STORAGE VOIDS AND COLONY SPACES MAY BE HANDED
PLACING BEES AGAINST WINTER PARTITION

THE BASIC PLAN

THE DECISION TO ADOPT LETALONE BEEKEEPING IS BEST WITH A NEW QUEEN EVERY
YEAR IN A HIVE WITH SUFFICIENT SPACE TO EXPAND INTO, BUT NOT TOO MUCH AS TO
COOL THE COLONY.

OPTION 1.

TAKE THE HONEY AT THE END OF JULY AND MAKE AN ARTIFICIAL SWARM WITH A
COLONY AT EACH END OF THE ZEST TO GO THROUGH THE WINTER. THE ARTIFICIAL
SWARM WILL NEED HEAVY FEEDING, THE RESIDUE LESS SO.

THE ADVANTAGES ARE:

A. TWO COLONIES TO OVERWINTER AS INSURANCE AGAINST THE LOSS OF ONE.

B. CAN BE RE-UNITED IN SPRING WHEN COLONY NEEDS TO BE STRONG.

C. IF BOTH COLONIES SURVIVE THE WINTER ONE CAN BE USED TO SEED A FURTHER
ZEST HIVE.

OPTION 2. (SEE DRAWING 8B)

MAKE NUCLEI FROM STRONG COLONIES EARLY IN THE SEASON TO USE BOTH ENDS OF
THE ZEST IF NOT ALREADY USED.

BEE ENTRY AND
VENTILATION AS
REQUIRED

———————— WOOD

———————— INSULATI[

8A

THE "ZEST" NUCLEUS MANAGEMENT PLAN

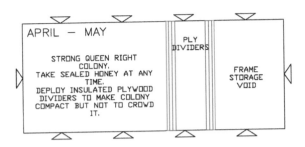

APRIL — MAY

STRONG QUEEN RIGHT COLONY.
TAKE SEALED HONEY AT ANY TIME.
DEPLOY INSULATED PLYWOOD DIVIDERS TO MAKE COLONY COMPACT BUT NOT TO CROWD IT.

PLY DIVIDERS

FRAME STORAGE VOID

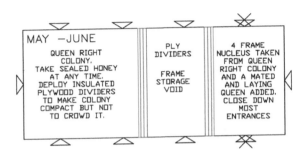

MAY —JUNE

QUEEN RIGHT COLONY.
TAKE SEALED HONEY AT ANY TIME.
DEPLOY INSULATED PLYWOOD DIVIDERS TO MAKE COLONY COMPACT BUT NOT TO CROWD IT.

PLY DIVIDERS

FRAME STORAGE VOID

4 FRAME NUCLEUS TAKEN FROM QUEEN RIGHT COLONY AND A MATED AND LAYING QUEEN ADDED. CLOSE DOWN MOST ENTRANCES

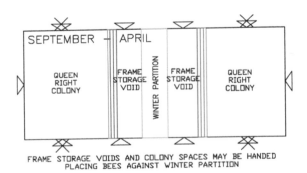

SEPTEMBER — APRIL

QUEEN RIGHT COLONY

FRAME STORAGE VOID

WINTER PARTITION

FRAME STORAGE VOID

QUEEN RIGHT COLONY

FRAME STORAGE VOIDS AND COLONY SPACES MAY BE HANDED
PLACING BEES AGAINST WINTER PARTITION

THE BASIC PLAN
THE DECISION TO ADOPT LETALONE BEEKEEPING IS BEST WITH A NEW QUEEN EVERY YEAR IN A HIVE WITH SUFFICIENT SPACE TO EXPAND INTO, BUT NOT TOO MUCH AS TO COOL THE COLONY.
OPTION 2.
MAKE NUCLEI FROM STRONG COLONIES EARLY IN THE SEASON TO USE BOTH ENDS OF THE ZEST IF NOT ALREADY USED.
FOR OPTION 1, SEE DRAWING 8A.

BEE ENTRY AND VENTILATION AS REQUIRED

————————— WOOD

————————— INSULATION

8B

The ZEST collapsable bamboo frame

THE FRAME

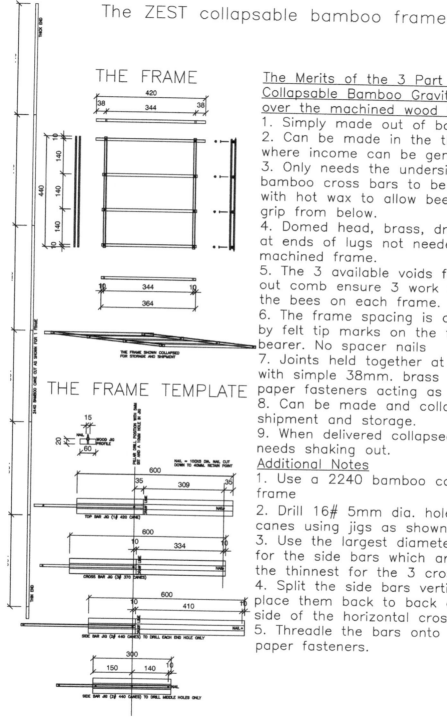

THE FRAME TEMPLATE

The Merits of the 3 Part ZEST Collapsable Bamboo Gravity Frame over the machined wood version.

1. Simply made out of bamboo.
2. Can be made in the third world where income can be generated.
3. Only needs the undersides of the bamboo cross bars to be brushed with hot wax to allow bees a firm grip from below.
4. Domed head, brass, drawing pins at ends of lugs not needed as machined frame.
5. The 3 available voids for drawing out comb ensure 3 work stations for the bees on each frame.
6. The frame spacing is determined by felt tip marks on the frame bearer. No spacer nails
7. Joints held together at junctions with simple 38mm. brass plated paper fasteners acting as split pins.
8. Can be made and collapsed for shipment and storage.
9. When delivered collapsed it merely needs shaking out.

Additional Notes

1. Use a 2240 bamboo cane for 1 frame
2. Drill 16# 5mm dia. holes in the canes using jigs as shown.
3. Use the largest diameter canes for the side bars which are split and the thinnest for the 3 cross bars.
4. Split the side bars vertically and place them back to back on each side of the horizontal cross bars.
5. Threadle the bars onto the brass paper fasteners.

9

The ZEST stapled wood frame

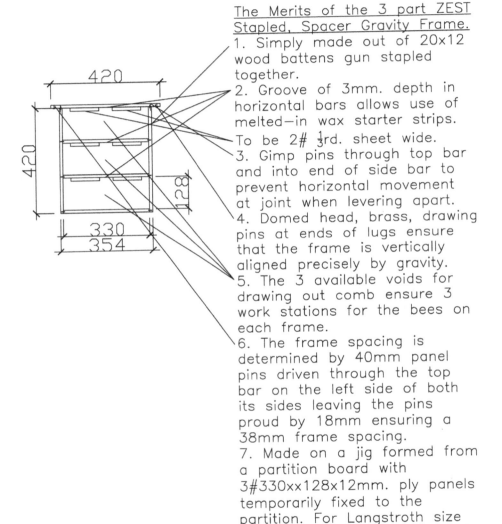

The Merits of the 3 part ZEST
Stapled, Spacer Gravity Frame.
1. Simply made out of 20x12
wood battens gun stapled
together.
2. Groove of 3mm. depth in
horizontal bars allows use of
melted−in wax starter strips.
To be 2# $\frac{1}{3}$rd. sheet wide.
3. Gimp pins through top bar
and into end of side bar to
prevent horizontal movement
at joint when levering apart.
4. Domed head, brass, drawing
pins at ends of lugs ensure
that the frame is vertically
aligned precisely by gravity.
5. The 3 available voids for
drawing out comb ensure 3
work stations for the bees on
each frame.
6. The frame spacing is
determined by 40mm panel
pins driven through the top
bar on the left side of both
its sides leaving the pins
proud by 18mm ensuring a
38mm frame spacing.
7. Made on a jig formed from
a partition board with
3#330xx128x12mm. ply panels
temporarily fixed to the
partition. For Langstroth size
frames increase void widths to
420. Top bar increases to 510

10

B.S. NATIONAL TOP BEARER FRAME
(LANGSTROTH TO BE 90MM WIDER)

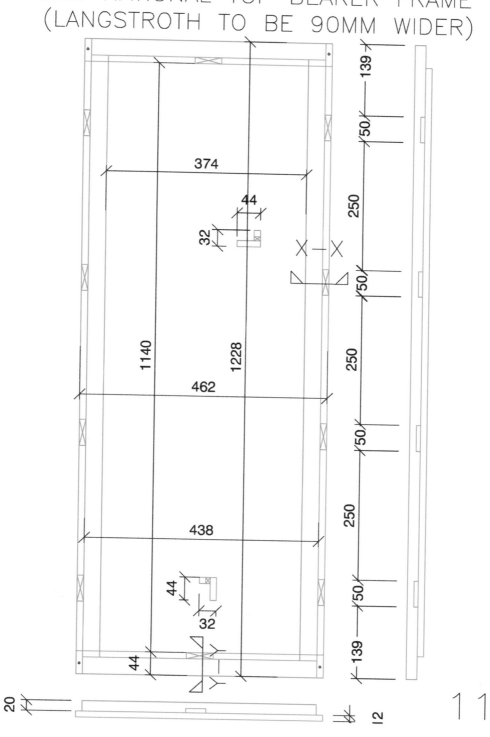

11

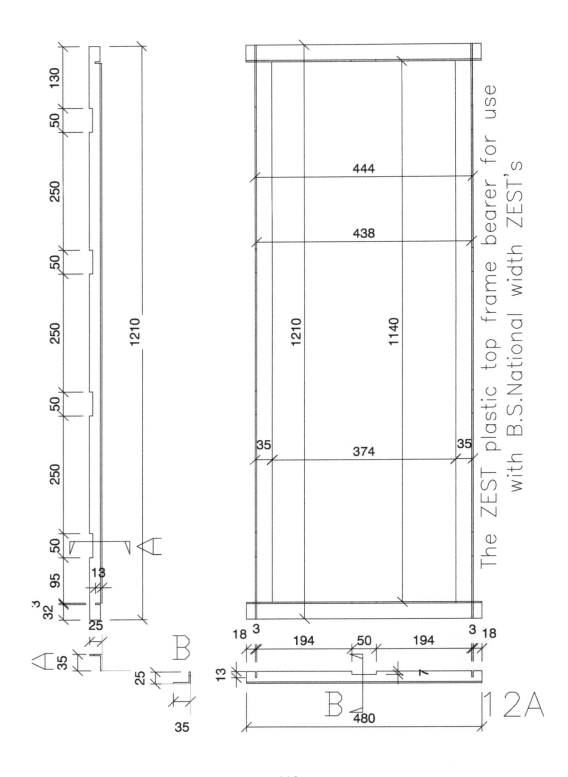

The ZEST plastic top frame bearer for use with B.S.National width ZEST's

12A

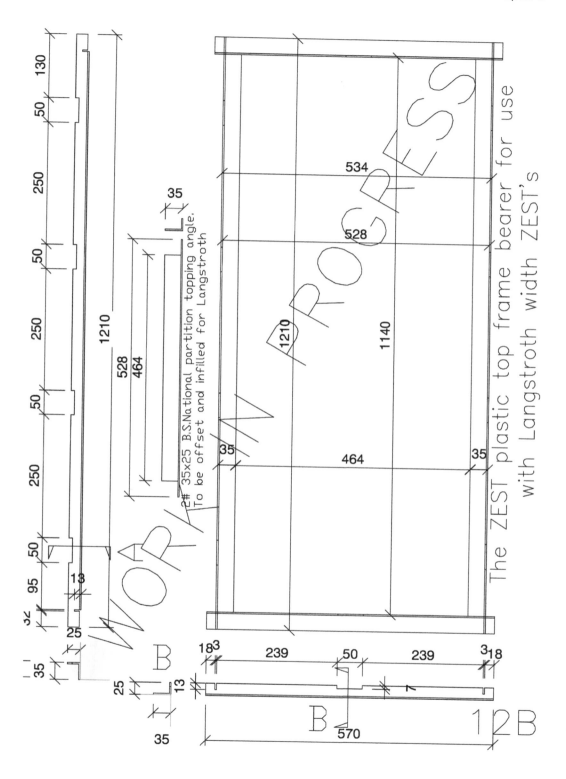

35x25 B.S.National partition topping angle,
To be offset and infilled for Langstroth

The ZEST plastic top frame bearer for use
with Langstroth width ZEST's

9 781908 904690